Systems thinking

Nic J. T. A. Kramer
Eindhoven University of Technology

Jacob de Smit
Interfaculty for Business administration, Delft-Rotterdam

Systems thinking
Concepts and notions

Martinus Nijhoff Social Sciences Division
Leiden 1977

Translation of the Dutch edition *Systeemdenken*, H. E. Stenfert Kroese B.V., Leiden 1974.

ISBN 90 207 0587 3

Printed in the Netherlands

Foreword

There is no generally accepted, clearly delineated body of knowledge concerning systems thinking. The multiplicity of thinking is well illustrated by the various names such as: (general) systems theory, systems thinking, systems approach, systems analysis, systems synthesis, systems engineering, etc. These terms refer to various fields of knowledge that either overlap or are completely different. For this reason we consider it useful to try to develop a common language, a common set of concepts.

In this book we have tried to launch such a common language. We shall try to develop a set of coherent concepts and notions. We have consciously tried to make the minimum use of mathematical or logical symbols in our descriptions and definitions. This promotes more positive access to the concepts. We think that the language of the formal sciences, mathematics, can only be partly of use to us in considering the application of systems thinking in complex empirical situations. Our set of concepts is based on various descriptions known from the literature. In order to explain the concepts and ideas as clearly as possible, we have illustrated them with examples from various academic fields such as sociology, psychology, business, management, economics, technology and the natural sciences. In the main, we have chosen relatively simple examples.

We feel that, after studying this book, the reader will be able to use the concepts and notions discussed in his own practical situation. In this context a closer study of some specific area of systems thinking may be necessary. For this purpose we have added references to each chapter relating to the subjects dealt with in that chapter. These serve as a guide for closer study. For us, this list of literature has served as a reference for the chapter concerned.

This book has been written as an initial introduction to systems thinking. It is intended primarily for those who want to learn about this subject for the first time. We have in mind managers, professionals, administrators and others, who deal with problems in complex systems, such as organizations. By dealing with concepts and notions from systems thinking they will probably obtain more insight into the possible origins and solutions of these problems than with a traditional analytical approach.

We also have in mind academics and students who can use this book in their studies and training as an introduction to systems thinking.

We are grateful to R. F. Hendriksen who made a valuable contribution to the initial draft of various chapters of the book.

Finally, as regards each another's contribution to this work, the entire text is the result of complete collaboration, and it is impossible to make any distinction. We want to dismiss any inference readers might make from the order of our names on the jacket.

Zadeh and Desoer suggested in the foreword to 'Linear System Theory' that people whose names begin with a letter early in the alphabet are generally at an advantage where the order is alphabetical. For this reason they decided the order of their names by tossing a coin, and 'Z' won. We have kept the customary alphabetical order.

Delft, The Netherlands
august 1974

Contents

CHAPTER 1

INTRODUCTION

1.1. SYSTEMS

An attempt to arrange present popular concepts and catchwords in rank order would certainly put the word 'system' high in the list. This concept is used in almost all sciences and has penetrated into normal everyday language and into the jargon of the mass media. Very many publications, conferences, symposia and university courses are devoted to systems, systems science, systems thinking, systems design, systems analysis, systems engineering, and so on.

As a result of its application to different fields, much confusion has been created around the concept of the system. This book will concentrate on systems thinking or systems science. What do we want with systems thinking? How was systems thinking created?

It was found in various academic disciplines that the problems, the objects of research, were becoming increasingly complex owing to progress in the respective sciences [1]. The traditional method of research was proving itself less and less adequate for dealing with these increasingly complex problems. Perhaps some examples will illustrate this.

The evolution of engineering from energy supply to control theory [2] has led to computers and automation. Automated machines have appeared, from the simple thermostat to the automatically piloted and self-correcting rockets of today. In the past, a steam engine, an automobile or a radio receiver would each lie within the range of knowledge, the competence, of an engineer who had been trained in these areas. But when we speak of spacecraft, large airplanes or ships, the mechanical, electrical and other systems prove so dependent on one another, that the proper functioning of the system as a whole is largely determined by these interrelationships. A new approach becomes essential: the systems approach. In this, it is essential to consider the system as a whole consisting of interdependent elements.

A second example is the development of biology. This first went in the direction of molecular biology, with such achievements as an understanding of the genetic code. Yet certain phenomena could not be explained by molecular biology. One of the great founders of general systems theory, Ludwig von Bertalanffy, pointed this out as early as 1928 [3]. From that time on this writer stressed the need for a fundamentally different approach, which he called 'organismic' biology.

A third example can be found in psychology, where originally a basic concept was used, sometimes called the 'robot' model. Behaviour was explained with the aid of a mechanistic stimulus-response model. These techniques, however, did not provide an adequate explanation of some aspects of human behaviour. This subsequently led to the development of 'Gestalt-'psychology, in which one proceeds from the whole (Gestalt) [4].

In very many social problems it has been found that the results of studies will improve with a holistic approach, in which emphasis is put mainly on the interrelationship of individual parts. Such problems are those, for instance, of air and water pollution, traffic congestion, and urban planning. A clear and well-known example of such an approach is 'The Limits to Growth, a Report for the Club of Rome, Project on the Predicament of Mankind' [5], which puts the emphasis on the mutual relationships between social class, population problems, food problems and so on.

1.2. HISTORICAL BACKGROUND[1]

Looking at the historical development of systems approach or systems thinking, we observe a striking parallel development in various sciences up to about 1950. If we first take the line of development in the humanities, and the natural sciences such as psychology and biology, we recognize the following important points.

In 1924 the German physicist Köhler [6], in his book on physical '*Gestalten*', gave the first impulse towards what could be called a general systems theory. He dealt with 'Gestalten' (wholes) from physics, but did not succeed in working out the problem in general terms. In a later article (1927) [7] he made another attempt.

He tried to find general qualities of organic systems by analogy with inorganic systems. The solution of this problem was the theory of open systems. In 1925 Lotka [8] published the first work on this subject, introducing the concept of 'open systems'. He observed systems interacting with their environment.

Then came the transition from molecular biology to 'organismic' biology; the best known names are Ludwig von Bertalanffy, who wrote on this in 1925-28 [3]; Whitehead, who first wrote on 'organic mechanism' in 1925 [9] and Cannon, who dealt with the concept of homeostasis in 1929 and 1932 [10]. The wide range of these concepts was only generally recognized around 1965, when a number of American biologists brought them back into the limelight.

The theory of open systems was further developed in biology. Via theoretical biology, the foundations of which were laid by Von Bertalanffy [11] in 1932, this led to the first attempt at a general systems theory after the Second World War. The essence of the general systems theory, expressed as: 'the whole is more than the sum of its parts', was at first not accepted by many people. In order

1. In this section we rely partly on the description of the history of general systems theory as given by Von Bertalanffy [1].

to gain wider support for their ideas, the economist Boulding, the mathematical biologist Rapoport, the physiologist Gerard and the biologist Von Bertalanffy founded the 'Society of General Systems Theory'. In 1957 its name was changed to the less pretentious 'Society for General Systems Research'. Under its auspices, contributions to general systems theory in various sciences are published in its 'Yearbooks of the Society for General Systems Research' [12], which have been published since 1956.

Parallel to these developments, there were also contributions before 1950 in cybernetics (developed largely from control theory), thermodynamics and information theory. In this area Szilard was in the forefront with his observations on the concept of entropy (1929) [13]. This was the first time the relationship between entropy and information and their opposite effect had been pointed out. The most important impulse to the development of General Systems Theory from these movements was in the period around the Second World War, given by Norbert Wiener (with Bigelow and Rosenblueth) in research into the design of automatic gun-firing control systems. This led in 1948 to Wiener's book 'Cybernetics' [14]. Its most important elements were the concepts of 'feedback' and 'homeostasis' and their formalization. In 1949 the information concept was formalized further in Shannon and Weaver's mathematical information theory [15].

It is noteworthy that the cyberneticians have in part joined the Society for General Systems Research and have come to regard cybernetics as a part of general systems theory, but also that societies of cyberneticians have widened their approach toward systems theory, as in the 'World Organization of General Systems and Cybernetics' [16].

There were several forerunners among the scientists who applied systems theory or systems thinking to management or corporate management problems. Stafford Beer, whose work was based mainly on cybernetics, published 'Cybernetics and Management' in 1959 [17]. Russell L. Ackoff and C. West Churchman [18], [19], tried to apply the systems approach to management problems via operations research, while Herbert A. Simon [20] also attempted this, as one of the first researchers on organization.

1.3. OBJECTIVE OF SYSTEMS THINKING

As already stated, a systems approach or systems thinking is a means of tackling problems, a methodology. A way of approaching problems which follows from two basic premises:

1. Reality is regarded in terms of wholes, 'Gestalten'.
2. The environment is regarded as essential, systems as in inter-action with the environment, as open systems.

What, then, is the goal of this General System(s) Theory (G.S.T.)? We quote the founder of the G.S.T., Ludwig von Bertalanffy, who with five main propositions defines the objective of the G.S.T. in his article 'General System Theory', published in 1956 in the first issue of the 'Yearbooks' [21]. These five points are:

1. 'There is a general tendency towards integration in the various sciences, natural and social'.
2. 'Such integration seems to be centered in a general theory of systems'.
3. 'Such theory may be an important means for aiming at exact theory in the non-physical fields of science'.
4. 'Developing unifying principles running 'vertically' through the universe of the individual sciences, this theory brings us near to the goal of the unity of science'.
5. 'This can lead to a much needed integration of scientific education'.

In the literature the title G.S.T. is used for General System Theory as well as General Systems Theory. In General System Theory one finds especially the ideas of Von Bertalanffy who, as stated above, examines in particular the 'unity of science' and believes he finds this in a General System: a general theory of systems valid for all systems. General System*s* Theory endeavours mainly to develop a 'system of systems': an all-encompassing, classifying, and relating theory concerning systems. We regard the two – General System Theory and General System*s* Theory – as well as their related variants, indicated by 'modern systems research', 'systems science', 'systems thinking', 'systems analysis', as parts of one field of scientific endeavour.

Our premise is that systems thinking has the following functions:

1. The development of a common language which makes it possible for scientists of different disciplines to communicate with one another: in other words a multidisciplinary means of communication. (A quotation from Boulding [22] has become famous in this respect: 'One wonders sometimes if science will not grind to a stop in an assemblage of walled-in hermits, each mumbling to himself words in a private language that only he can understand', in which he indicated how much he senses the need for a common language.)
2. To give an insight into the methodology inherent in the systems approach, and which is in fact the approach from the outside to the inside, a holistic approach.

1.4. SYSTEMS THINKING AND THE THEORIES OF ORGANIZATION AND MANAGEMENT

Systems thinking can play an important role in the development of theories of organization and management for the following reason.

These theories are part of a developing science devoted to complex organizations. After all, the organization as an object is the central point of study. This object, in order to be studied most fruitfully, must be approached as integrally as possible, with the relevant monodisciplines adding their share of knowledge regarding the many different aspects we have to deal with. We can then talk about a multidisciplinary approach. An approach based on systems thinking will make such a multidisciplinary approach simpler.

1. The approach from as many aspects as possible (multidisciplinary) can bring improved results by starting from a system approach. Right away this will reduce the possibility of some aspects being stressed more than others.
2. In the integration of these aspects, the presence of a common language can be a great help. It should be noted that such common languages already exist. Mathematics is one of them. But the language of mathematics will not always suffice for solving management problems since the formalization of empirical phenomena in organization and management often creates particularly difficult problems.

In the thinking about organizations more than anywhere else, we find the strong tendency of the monodisciplines to emphasize different aspects (the economist views a corporation differently from the engineer or the sociologist). For this reason we have chosen to focus on the development of this common language.

The second point of departure is somewhat more ambitious. The complex problems arising in organizations often require a different methodology: one that is based on the whole, a systems approach. In this book we can only acquaint the reader brieflly with this methodology.

The systems approach is already found in various parts of managerial sciences, such as organization theory, where the organization is regarded as an integrated complex of interdependent parts capable of interacting sensitively and correctly with one another and with their environment. A second example is that of business administration, where bookkeeping has developed towards

management information systems [23]. As a final example we may
mention the construction of integral corporate models, where a
system approach is essential in order to evaluate the behaviour of a
corporation as a whole or that of its parts.

1.5. SUBJECTS FOR DISCUSSION

In this first chapter we have outlined the place and development of
systems thinking. In the second chapter we shall discuss the various
concepts by describing the systems themselves.

In chapter 3 we shall discuss the surroundings of systems, what
lies outside them: their environment. We shall examine the criteria
for differentiating between system and environment and the influence this environment has on the system under consideration.

In chapter 4 we shall examine the behaviour of systems and
characteristics of this behaviour. We shall look, for example, at
such important characteristics as the equifinality and stability of
systems.

Chapter 5 is devoted to communication and information. We also
discuss entropy and the relationships between entropy and information. Reasoning by analogy about such processes, in systems from
the natural and managerial sciences, such as are found in the
literature on organizations, is also discussed. In these chapters we
shall present, so to speak, a foundation for an explanatory desk
dictionary of the language of systems.

We have tried to interrelate the concepts and to clarify them with
examples. The descriptions we have chosen should be looked upon
as aids to starting discussion about these concepts. We do not claim
our descriptions to be the best possible ones.

Chapter 6 will go into the description of systems and the related
possibilities of enquiring into them. We shall discuss models, the
concept of the model and different kinds of models. We shall also
deal with model construction, quantitive descriptions of systems
and descriptions of systems in terms of input and output. Finally

we shall devote a paragraph to Boulding's classification of systems [22], to give us an insight into the position of systems research. We pay attention to this because the systems occurring in reality are almost invariably too complex to describe adequately, to comprehend, or to work with. We always form a model of such a system for ourselves, whether this model is a description, a representation, or a set of equations. Rosenblueth and Wiener [24] comment on this: 'No substantial part of the universe is so simple that it can be grasped and controlled without abstraction. Abstraction consists in replacing the part of the universe under consideration by a model of similar but simpler structure. Models, formal or intellectual on the one hand, or material on the other, are thus a central necessity of scientific procedure'.

In the last chapter is discussed a special class of systems, the cybernetic systems. These receive our attention because they have special characteristics, such as goal-directed behaviour, information flows, or dynamic behaviour, which are also of great importance in studying more complex systems. Through the directed influencing of other systems they are able to control these systems. Control is such a universal and complex phenomenon in our world that models based on these cybernetic systems are about the most complex ones available for studying systems that are even more complex in origin.

1.6. REFERENCES

1. Bertalanffy, L. von, *General System Theory*, New York 1968.
2. Boiten, R. G., 'Cybernetica en samenleving', in *Arbeid op de tweesprong*, The Hague 1965.
3. Bertalanffy, L. von, *Kritische Theorie der Formbildung*, Berlin, Borntraeger 1928.
4. Wertheimer, M., 'Untersuchungen zur Lehre von der Gestalt', *Psychol. Forsch.* 4, 1923.
5. Meadows, D., et al., *The limits to growth, a report for the Club of Rome project on the predicament of mankind*, New York 1972.
6. Köhler, W., *Die physischen Gestalten in Ruhe und im stationären Zustand*, Erlangen 1924.
7. Köhler, W., 'Zum Problem der Regulation', *Roux's Arch.* 112, 1927.

8. Lotka, A. J., *Elements of Physical Biology*, New York, Dover (1925), 1956.
9. Whitehead, A. N., *Science and the modern World*, Lowell Lectures 1925, New York 1953.
10a. Cannon, W. B., 'Organization for physiological homeostasis', *Physiological Review 9*, 1929.
10b. Cannon, W. B., *The wisdom of the body*, New York 1932.
11. Bertalanffy, L. von, *Theoretische Biologie*, I + II, Berlin, Borntraeger 1932, 1942.
12. Yearbooks of the Society for General Systems Research, *General Systems* I-XXI, 1956-1976.
13. Szilard, L., 'Über die Entropieverminderung in einem thermodynamischen System bei Eingriffen intelligenter Wesen', *Z. für Physik*, 1929, p. 840.
14. Wiener, N., *Cybernetics*, New York 1948.
15. Shannon, C. E. and W. Weaver, *The mathematical theory of communication*, Urbana 1949.
16. *Proceedings of the conferences of the World Organization for General Systems and Cybernetics*, J. Rose, ed., 1969/1972/1975.
17. Beer, S., *Cybernetics and management*, London 1959.
18. Ackoff, R. L., 'Systems, organizations and interdisciplinary research', *General Systems* V, 1960.
19. Churchman, C. W., et al., *Introduction to operations research*, I, New York 1957.
20. Simon, H. A., *Models of man*, New York 1957.
21. Bertalanffy, L. von, 'General system theory', *General Systems* I, 1956.
22. Boulding, K. E., 'General systems theory – The skeleton of science', *General Systems* I, 1956.
23. Murdick, R. G. and J. E. Ross, *Information systems for modern management*, Englewood Cliffs, N. J., 1971.
24. Rosenblueth, A. and N. Wiener, 'The role of models in science', *Phil. of Sci.* 12, 1945.

CHAPTER 2

SYSTEM

2.1. INTRODUCTION [25], [26], [27]

In this chapter we present a set of concepts which are as coherent as possible and which will enable us to describe the inner structure and the behaviour of systems. Most of these concepts are already used in many branches of science. The interpretation often varies. In studying complex problems which require a multidisciplinary approach, this can easily cause semantic confusion. Communication between people in different disciplines, which is a prerequisite for the optimal solution of a given problem, is thus made difficult if not impossible.

We present a coherent set of concepts of such generality that it becomes possible to build bridges between different disciplines. This can be done by using the same concepts in these different disciplines. Through the use of a single language there will be more chance of 'cross fertilisation' between disciplines, which will then help to demolish the language barrier.

In order to avoid accentuating the gap in formalism, we have tried to avoid formal mathematical definitions. So doing, we hope to obtain the indispensable communication between the sciences, unhampered by differences in terminology. Therefore we begin with the following idea of a system in the empirical world around us (fig. 1).

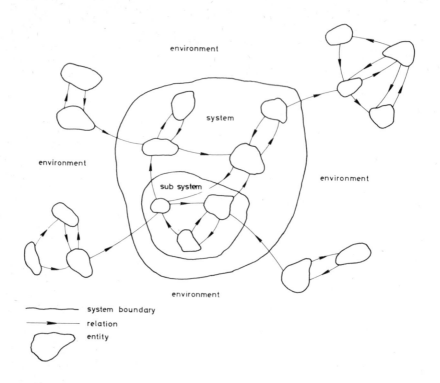

Figure 1. Representation of a system.

As already mentioned, this chapter will deal with the inner structure and the behaviour of the system. We shall deal with its description by looking at its structure, and with its dynamics by looking at the state resulting from a certain behaviour of a system.

Our method of presenting these concepts should be regarded in the light of our objective: the cultivation of understanding of how these concepts are applied or are applicable in various sciences.

Of primary importance is the development of a common language. Insights obtained in mono-disciplines, if formulated in this language, become easier to transfer to other disciplines. In this way multi-disciplines can be created and can develop into an interdiscipline, which can make possible the necessary interdisciplinary approach.

We are aware that such an interdisciplinary approach requires a re-orientation of many concepts which come forth from the basic disciplines, so that the specific interpretation of these concepts can still change in the basic disciplines.

2.2. SYSTEM AND AGGREGATE

In the empirical world around us we observe various sets of entities, objects or things, such as: a group of people; a combination of people and machines; a biological organism; a solar system; libraries; automobiles; a vase of flowers. We can consider these in two basically different ways: 'as a system' or 'as an aggregate'. The difference between these two is clearly expressed by Angyal [28] when he says: 'in a system it is significant that the parts are arranged' and 'in an aggregate it is significant that the parts are added'. The difference lies in the 'relationship to the whole' (see 2.4), which is present in a system but not in an aggregate. The following examples make this clear: in the English language we have the letters (entities) m, a, e and t. Putting these together in a random fashion gives us meaningless combinations such as maet (see 5.2.3). These combinations we call aggregates. There are, however, combinations such as meat, team, which do possess the feature of meaningful 'arrangement'. Within the English language we could regard these as systems.

We have now reached a very important aspect of systems definition. We see from this example that it is somewhat arbitrary whether we call something a system or not. In another language a combination such as *maet* may very well have a meaning, and would then be considered a system. Therefore we suggest that a system can only be defined as an object of research or observation [29], [30]. One tends to say: we look at this or that 'as a system'. This means that we can label certain known objects as systems if they fulfil a system's requirements.

In order to be able to judge this, we first have to define a system. It is:

'a set of interrelated entities, of which no subset is unrelated to any other subset'.

This means that a system as a whole displays properties which none of its parts or subsets has. It also means that every entity in the system is either directly or indirectly related to every other entity in it. An aggregate, then, is a set of entities which may perhaps be partly interrelated, but in which at least one entity or subset of entities is unrelated to the complementary set of entities.

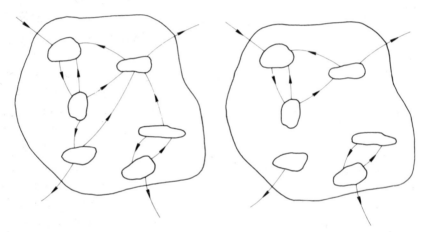

Figure 2. System.   *Figure 3.* Aggregate.

2.3. ENTITY

Entities are the elements or parts of a system. We consider these as primitive terms, dispensing with the need for definition. In order to illustrate the meaning of an entity, we can use the definition in Webster's Dictionary [31]:

'An entity is something that has objective or physical reality and distinction of being and character'.

We purposely use the concept of entity in our definition of a system, because we consider it a more neutral notion than a term such as element, which invokes associations with the smallest part that still possesses certain qualities, or a term such as part, which does the opposite. Other possible concepts might be object [32], or thing.

Depending on the kind of system, entities can be interpreted as: mathematical quantities, airplanes, signs, individuals, and so on.

We can attribute certain properties or attributes [33] to entities. By means of these, we can distinguish the entity (or the system) from others. One can attribute certain values to the properties by measuring them. The properties of the observed entities define, in fact, the degree of their recognizability. Entities without properties cannot be observed.

2.4. RELATION

We could regard the concept of the relation as a primitive term, similarly to the concept of the entity. But there are some misunderstandings concerning the concept of the relation, mainly because of the complex effect different kinds of relations have on the behaviour of the system as a whole. We therefore feel that a broader discussion of this concept is appropriate.

We define a relation as:

> 'the way in which two or more entities are dependent on each other'.

It must be added that this definition uses entities more in the sense of their properties (attributes). Between entities to which no properties are attributed there can be no interrelationships; such entities cannot be observed.

What is the extent of the given definition of a relation? A relation links the properties of various entities. The (mathematical) description of this linkage readily assumes the form of a (mathematical) function. Such a linkage also provides scope for interaction. The behaviour of the various entities is then no longer independent. A relation exists '*if a change in a property of one entity results in a change in a property of another entity*'.

The existence of a certain relation imposes a constraint on a system's possible modes of behaviour. This is a direct consequence of the interdependence resulting from the existence of a relation.

Hence there is interdependence in the location of two objects, for instance two spheres joined together by a rod.

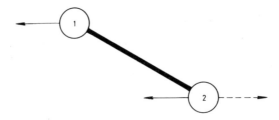

Figure 4. Two spheres joined together by a rod.

If sphere 1 is moved to the left, sphere 2 will not be able to move to the right, provided the rod is rigid. A constraint is created upon the behaviour of sphere 2, but of sphere 1 as well.

These constraints upon the behaviour of entities in a system as a result of the relation between these entities are governed by the nature of the relation. If we could say anything a priori about the nature of the relationships in question, the explanation of the behaviour of entities or systems would probably be made easier. For this reason it is useful, when dealing with the concept of the relation, to distinguish between the 'relation of an entity toward the whole' and 'the relations of an entity toward other entities'. This distinction agrees with those of Shchedrovitzky [34] and Angyal [28].

The 'relation to the whole', the one-whole relation, is essential in a system. Without it, a phenomenon cannot be viewed as a system. But in practice it is often difficult to differentiate because a relation

may have both, one-whole and one-to-one characteristics. Therefore the concepts are ideal types, characterizing both ends of a scale.
They are useful in thinking about the effect of relations on systems
and the behaviour of systems. To make the distinction credible, we
shall illustrate it with an example [34]. Let us take a board with 4
holes in it. In each hole there is a marble (see fig. 5).

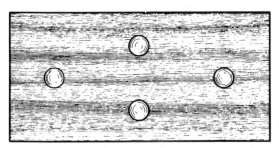

Figure 5. Configuration 1.

This configuration could be described with a holistic approach (an
approach based on the study of the system as an organized object
[28]). In such a description we use the general characteristics of a
rhomb (from plane geometry) whose properties are determined by
its being equilateral and by the size of a single angle. This completely
describes and defines the configuration; it is a system because if we
were to remove one of the marbles, its character would change completely. We would have an isosceles triangle, which has characteristics different from those of a rhomb.

It is important to note that changing the position of one marble
does not change the position of the others, though there is a change
in the whole. A system with this characteristic is called 'an organization with relations' (Shchedrovitzky [34]), while Angyal speaks of
the 'relation to the whole' and 'connections to the superordinate
system' [28]. Hence in this example the system is determined by the
relations to the whole.

Let us take the same example, but now with springs between the
marbles.

While at rest, the marbles are in the same position as in the first example. Moving one of the marbles now has a totally different effect. All the marbles will move along to some extent, and owing to the tension of the springs a new configuration will emerge which, however, largely resembles the old one. In principle, a new rhomb will be formed if the springs are pulled to the same extent. A change has occurred in the whole in which, unlike the first example, the position of all the elements has changed. Certain qualities of the whole have remained, however. They are not only related to one another, but are also connected to one another.

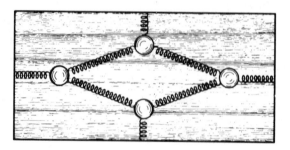

Figure 6. Configuration 2.

Shchedrovitzky [34] calls a system with such relations 'a system of structural connections'. Angyal speaks of 'two term relations' [28]. This system is marked especially by the nature of the 'connections', by the interrelations between the entities and not by the relations to the whole.

Consequently, in a 'organization with relations' a change in one of the 'relations' implies complete change in the whole, while this is not necessarily so in the case of a 'system of structural connections', since the relations to the whole play either no role at all or only a minor one. We shall revert in greater detail to the concept of 'relation to the whole' and the concept derived from this of: 'positional value' of an entity in a system when we deal with the concept of structure (2.5).

There are two distinct ways [33] of establishing the existence of a relation:

1. *by experiment*: change property a of entity 1 and see if a change occurs in a property of entity 2;
2. *by recording*: record the changes in property a of entity 1 and property b of entity 2 and check if there is a connection (function) which can describe changes in a and b at the same time.

There is, however, a possibility of a relation observed at a certain moment no longer existing at another moment. The connection observed earlier may be of an incidental nature. In order to establish the existence of a relation during a certain time with a certain validity, the observations must be made for a prolonged period of time. This period would be specifically defined.

We would again point out that the researcher's standpoint is of very great importance. It may mean that some relations are not taken into account while others, of importance to the problem he wants to solve, are disregarded. In general, we can assume or point out relations for each group of entities. The question, however, whether relations exist of importance to a given problem must be answered by the researcher. Obviously, the choice the researcher has to make of the relations he will study and those he will disregard is extremely important, since in this way he makes his initial choice with regard to the system in question. He must therefore state his choice explicitly. It will be influenced by a number of factors such as his objective and his perception of the situation. It is reflected in the choice of the sub and aspect systems considered and the resolution level (2.7).

2.5. STRUCTURE

When we concern ourselves with relations between entities, a new concept emerges which is regularly used in this context: that of 'structure'. This deals mainly with the relations recognized in the system. The literature often defines the concept of structure as the whole of the relations (set of relations).

Let us now discuss the concept of 'structure' in more detail, based on the following three points [28]:

1. the set of relations;
2. the positional value;
3. the dimensional domain.

We can depict this as follows. Take a system:

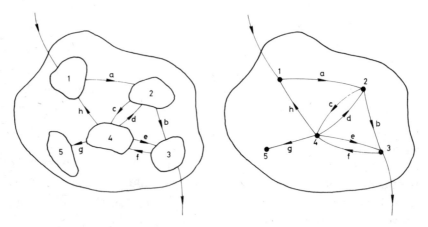

Figure 7. System and structure.

The set of relations (an atomistic assembly) consists of a, b, c, d, e, f, g, h; the positional value of the entities can be described (looking at the positions) as: 2 is to the right of 1,3 is to the right of 2, etc. The dimensional domain then gives content to the distance (if we are considering this relation) between two entities: a is 15 cm. and so on, and defines the structure in relation to other structures, for example as: relation 1-2 is easterly.

The first point indicates the existence of the relations between the entities. The second defines the arrangement of the entities, their position as related to one another within the system as a whole.

Relations between entities, and the arrangement of entities towards one another, presuppose the possibility of separating the entities. A system must have dimensions. The most obvious ones are space and time. This is expressed by the third characteristic of the concept of structure, the dimensional domain. This makes systems distinguishable from one another (multiplicity of entities). The following two examples will clarify this:

1. We can regard cities as entities of the system of Europe; roads as relations between the entities (cities). We can show all this on a map. But there are other possibilities too. We could, for example, list all the roads, Berlin - Brussels, Brussels - Paris, and so on. This is the set of relations. The positional value is indicated, for instance, by the distance from Brussels to Berlin is 5 and Brussels is at the left of the straight line between Berlin and Paris. The distance from Brussels to Paris is 2. The dimensional domain then gives the distance in miles and an indication of North.
2. A group of individuals can be regarded as a system. The set of relations consists of the various bonds between the individuals, such as love, work, family ties. The positional value indicates, for example, the relative dominance of one individual over the group; the dimensional domain the depth of the relation compared to other similar relations in the environment.

The concept of 'relation to the whole' and positional value associated with this makes it possible to distinguish between aggregate and system. Let us examine the following figure:

Figure 8 represents a system; figure 9 an aggregate (2.2). Both are located in the same environment. The structure of a system is determined by:

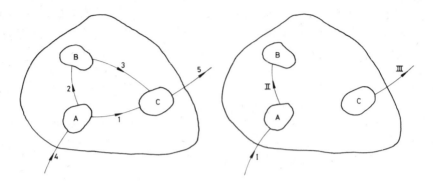

Figure 8. System.   *Figure 9.* Aggregate.

1. the set of relations: 1, 2, 3, 4, 5;
2. the positional value: A is below B, etc.;
3. the dimensional domain: location in relation to the environment.

In the aggregate we can speak of a set of relations (I, II, III) but not of the positional value of entities, since relations do not exist between all the entities. Hence there is no relation to the whole. The dimensional domain is the same as in the system.

The quotation from Angyal already mentioned is also pertinent in this context: 'in a system it is significant that the parts are arranged'; 'in an aggregate it is significant that the parts are added'.

It should be pointed out that some writers (including De Leeuw [33]) do not divide the notion of structure into the three characteristics mentioned above. They usually define structure as 'the set of relations'.

'Positional value' and the 'dimensional domain', however, are embodied as a relation within the set of relations, owing to a more formal approach to the concept of relation.

We have taken the liberty of adding these two characteristics more explicitly to the definition of structure because we think this clarifies the concept. Fundamentally, the two definitions do not differ.

2.6. STATE

So far we have defined the concept of system and discussed those of entity, relation and structure. We have dealt mainly with the framework within which the description of a system becomes possible. This description did not take into account the behaviour of the system. In order adequately to describe a system's dynamic behaviour we must introduce the concept of state.

To begin with, we can describe this concept in Russell L. Ackoff's words [26]. He says, '*the state of a system at a moment of time is the set of relevant properties which that system has at that time*'. We can now describe the dynamic behaviour of a system as follows. Take a system with an input and an output (see 3.3). The idea is to predict the output of the system if we know the input. In order to make this prediction possible, the concept of state is introduced. If we know the input and the state, then the output is fully defined.

Next we differentiate between 'deterministic' and 'stochastic' systems. In deterministic systems we have a complete knowledge of the future behaviour of the output as soon as we know input and state. In stochastic systems this is more difficult since probabilities play a role. By means of input and state we can calculate the probability of the occurrence of a certain output. The concept of state comprises, as it were, a part of the system's past. It can serve as a memory in which the relevant earlier history is stored so that, together with the input of the system, something can be said about the output.

The following definition of the concept of state, which emphasizes its dynamic character, may make it clearer:

> 'the state of a system, containing the information on the system's earlier history and its present condition, is necessary and sufficient for predicting the output or the probability of a certain output, given a certain input'.

A good example is banking. The payments made by a bank depend not only on the customer's payment orders but also on the balance he has in his account. The state is his balance. This is, in fact, a summary of the earlier history (deposits, payment orders, interest, etc.).

The researcher would like to make a priori statements about a system's state after studying its structure. This raises the question whether the concept of state can also be applied to entities and relations.

As regards entities, this question is easy to answer, since at a given moment any entity reveals a number of attributes (see 2.3). The combination of these attributes (including a memory) then forms the state of an entity.

As regards a relation, the question is harder to answer, since a relation is regarded as a link between attributes (state magnitudes) of entities. It is thus left undecided whether this link itself has certain properties. Yet in our opinion, in the case of relations there is also a state, because a relation has certain properties and can actively influence a system's behaviour. A relation has a dynamic character.

As an example: the relation between two entities may be a rod. As regards the relation between two entities it is essential to know whether the rod is rigid or elastic, since this quality is important to the extent of location interdependence resulting from the existence of the rod. If it is elastic, location dependence will not be as strong as if it is rigid.

This leads to the conclusion that a system's state is defined, among other things, by the set of states of the entities and the relations. We have now linked the concept of state to the properties of the relations and the entities as well, and have thus linked the concept of state to that of structure. Yet the state of the structure encompasses more than the set of states of the relations, since the positional value of the entities and the dimensional domain as part of the structure, also show the characteristics of a state.

One last remark. In the example of the rod as the relation between two entities, it might be suggested that the rod, which we call a relation, can be considered as an entity. We can go even further and say: could we study the system inversely? The original entities and relations then become relations and entities respectively. In the original system, we already attributed properties to both and these continue to exist in the inverse definition. Perhaps this will enable us to obtain a simpler explanation of the concept of state. If a normal and an inverse observation yield the same concept of state, then we can attribute properties to entities and relations in an equivalent way.

An obvious example is the so-called flow-chart used in planning techniques. These charts represent systems with spheres and arrows, as in fig. 10.

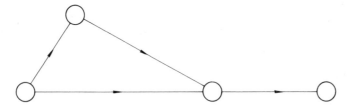

Figure 10. Flow-chart.

There are now two systems of presentation. One makes the activities occur in the spheres, the arrows transmitting the activities. The other makes the activities occur in the arrows, with the spheres transmitting the activities. Both methods yield the same result (there are complex one-to-one transformation rules), while their character is clearly inverse.

2.7. SUB AND ASPECT SYSTEMS

In our experience it is often useful, though less fundamental in studying systems, to differentiate between sub and aspect systems.

The clearest definition of a subsystem we have found is that in O. R. Young [27]. He says:

> 'a subsystem is an element or functional component of a larger system which fulfils the conditions of a system in itself, but which also plays a role in the operation of a larger system'.

That means in fact a part of the system which could be regarded as a system in itself. For us, this means that each entity could in fact be regarded as a subsystem. An entity is after all a part of the system to which we attribute certain properties. We do not split it up because, as researchers, we do not think this useful at a given moment. We stay at a certain 'resolution level' [35]. This concept will often be encountered in the theory or the application of the theory of 'integrated hierarchic subsystems' [36] to organizations. This theory endeavours to describe organizations as a tree of hierarchically ordered subsystems and stops at a certain resolution level depending on the problem. To make this clearer, let us examine the following three drawings.

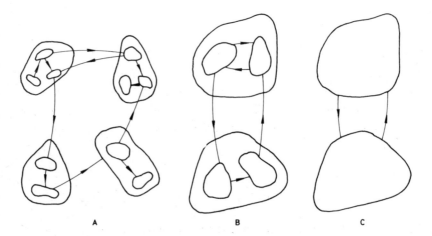

A   B   C

Figure 11. System at various resolution levels.

A, B, and C are all representations of the same system at different resolution levels. A has the highest level (most detailed), C the lowest. B is in between. When we study an entity or a system, we raise the resolution level. In general, such a part proves to contain a number of smaller, interrelated parts. This means that in the case of a subsystem we have in mind a recognizable part, separated from the other parts of the system, while all relations remain intact in that part itself and also in the interaction between such part and the rest of the system.

This contrasts with the 'aspect system'. Other authors [33] use the term 'partial system' for this, in which part of the relations unimportant to the aspect being studied are disregarded. The aspect system need not therefore function as a system itself.

The following is an example of the distinction: consider an organization. It comprises various kinds of relations. A production department can then be regarded as a subsystem, but the entire information system as an aspect system [37].

2.8. REFERENCES

25. Hall, A. D. and R. E. Fagen, 'Definition of a system', *General Systems* I, 1956.
26. Ackoff, R. L., 'Towards a system of systems concepts', *Man. Sci.* 17, July 1971.
27. Young, O. R., 'A survey of general systems theory, *General Systems* IX, 1964.
28. Angyal, A., 'A logic of systems', in: F. E. Emery (ed.), *Systems thinking*, Harmondsworth 1969.
29. Beer, S., *Cybernetics and management*, London 1959.
30. Wiener, N., *The human use of human beings*, New York 1954.
31. *Webster's Collegiate Dictionary*, 1970.
32. Leeuw, A. C. J. de, *Systeemleer* I, syllabus Eindhoven University of Technology, 1971.
33. Leeuw, A. C. J. de and W. Monhemius, *Methodologie en inleiding systeem-leer* II, syllabus Eindhoven University of Technology, 1973.
34. Shchedrovitzky, G. P., 'Methodological problems of system research', *General Systems* XI, 1966.

35. Klir, J. and M. Valach, *Cybernetic modelling*, London 1965.
36. Mesarovic, M. D., D. Macko and Y. Takahara, *Theory of hierarchical, multilevel systems*, New York 1970.
37. Maarschalk, C. G. D., 'The use of aspect systems in a general model for organizational structure and organizational control', in: B. van Rootselaar (ed.), *Annals of systems research*, vol. 1, Leiden 1971.

CHAPTER 3

ENVIRONMENT

3.1. INTRODUCTION

In this section we will discuss the terms 'system boundary', 'open and closed system', and 'environment'. These terms should be examined in their interrelationship in order to avoid misinterpretation.

It seems impossible to discuss the concept of the environment without having yet determined what belongs and does not belong to the system – in other words, without having determined the boundaries of the system. On the other hand, an environment is only relevant to open systems. Firstly, the difference between open and closed systems must be determined before being able to discuss the environment.

When discussing the concept of the system we conceived a system as built up of relations and entities. In looking at the complex world around us, it is difficult to decide what to call a system. We can regard the entire world as we know it as a system, but we can also select anything in it and describe that as a system. The number of relations and entities in the world around us is infinite. Looking around us we can obviously regard anything as a system, since everything is interrelated in some way. Hence we must select a group of entities and relations which we can call a system. We conceive of a system as a part of a larger whole. This larger whole then consists of the system and something else which we will refer to as its environment. We then envisage a system boundary between the system and its environment.

We speak of a closed system when the system does not interact or is not regarded as interacting with its environment. One could say that the environment does not exist in the case of such a system. In contrast to the closed system, we speak of an open system when a system interacts or is regarded as interacting with its environment.

3.2. SYSTEM BOUNDARY

In a superficial consideration of empirical systems we often tend to accept their physical limits as their boundaries, for example: man's skin, the wall round a factory, or the Iron Curtain. It is remarkable, however, that in scientific studies of physical systems the physical boundaries are often not taken as the boundary of the system. In practice, it is very difficult for the researcher to determine a system's boundary.

Let us give three examples of selecting a system and the consequent problems:

Firstly the Kingdom of the Netherlands: initially and superficially we might place the system boundary at the national frontiers. But there are then difficulties regarding other dimensions such as the height of the atmosphere over the country and the depth of the earth's crust under it, which may or may not be considered as part of the territory. What is an embassy in another country; is it a part of Dutch territory, and what about Dutch nationals in other countries?

The second example is that of a cyclist. There are several systems: the bicycle, the cyclist plus the bicycle, or the cyclist alone. Or we can regard the cyclist, the bicycle, the part of the road on which he is cycling, the ambient air, the friction and a headwind as part of the system.

One last example: a corporation. This may be a physical system with the work of the employees as the input of the system. If we select the factory walls as the system boundary, then at a regular hour on a regular working day a number of employees will be functioning as entities of the system. But the money in the bank, which is outside the factory walls, will not be part of the corporation as a system.

This perhaps illustrates that the selection or determination of a system boundary may involve problems. In practice our selection will depend on the purpose of the research. If, for instance, we want to check whether milk and butter are being stolen at a dairy plant (by comparing fat input and output) we shall take the factory walls as the system boundary.

It thus seems that a boundary can only be defined as something imaginary or conceptual inasmuch as everything within it can be considered as part of the system, and everything outside it as part of the environment. This means that the boundary is determined by a number of criteria which entities have to fulfil to be considered part of the system.

This contrasts with the entities failing to satisfy these criteria and essentially forming part of the environment. Many authors refer to boundary areas meaning (the boundary together with) the entities of the environment directly interacting with the system, and those directly interacting with the environment [38].

Hence a number of writers on management say: 'The primary task of management is to manage the boundary conditions of the enterprise' [39]. Apparently they consider this boundary area so important to the functioning of the enterprise that they elevate it to the most important objective of business management.

Some authors, including Ulrich [40], adopt a different approach for selecting or establishing a system boundary. They assume that the boundary of the system will be where relations are less concentrated than at other places. The higher concentration will then be within the system boundary.

In fig. 12 the boundary will be along the unbroken line. The

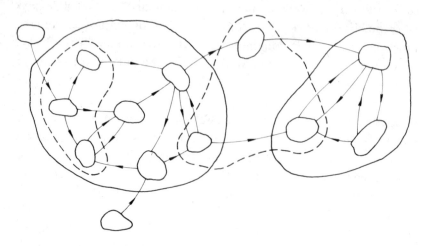

Figure 12. Selecting the system boundary.

selection of a boundary along the broken line is probably detrimental
to the whole of the system being examined. This approach is credible.
In reality, most organizations have many more internal relations
than external relations for example. To select the system boundary
right through such an organization is not a very apt solution.

3.3. OPEN AND CLOSED SYSTEMS

Physical systems known to us are by definition open systems. One
way or another they are related to their environments. All these
systems have a system boundary. However, the literature often
refers to 'closed systems'. A researcher considers a system as closed
(having no interaction with its environment) because he disregards
the relations that exist between the system and its environment.
They are not relevant to the problem, and he thinks that their influence on the system in question can be disregarded. This approach
often gives a reasonably quick insight into the system's internal
functioning at an early stage in research, but can also quickly lead

to incorrect interpretations of the phenomena, and to incorrect conclusions from these interpretations.

Two pencil and paper experiments are possible to arrive at a closed system. Firstly, we can move our system boundary outward and consider our system as something larger. At a certain moment it will pass beyond our comprehension, and we shall consider the system as closed (for example the universe).

Secondly, we can use the concept of the closed system as a paradigm. In our view this is a much more important and more concrete application of this concept. As stated earlier, it is useful for many studies to be able to abstract from interactions between the environment and the system. Examples are the closed economy model [41], proceeding from the assumption that no money moves beyond a country's frontiers (for didactical reasons this model is often used in study programmes); or models from thermodynamics [42], where a system is considered as closed when no matter is exchanged across the boundary line (this simplifies the study of energy exchanges). It is also an example of a system where the observer decides when to call the system open or closed.

According to Von Bertalanffy, a closed system is defined as follows [43]:

'closed systems are systems which are considered to be isolated from the environment'.

We can describe open systems as:

'a system is open when there is a set of entities which does not belong to the system but influences or is influenced by its state'.

If a system is regarded as open, the state of the system and that of the environment influence each other (see definition above): the system interacts with its environment. Hence there are relations between the system and the environment whereby at least one entity in the system influences the state of an entity in the environment or vice versa. There is not necessarily any interaction upon the state. For example, weather conditions influence man's state, but as far as we know, man's state does not affect weather conditions.

3.4. ENVIRONMENT

The foregoing might lead to the conclusion that anything not belonging to the system in question might be called its environment. But this definition is not always uniformly functional. We can distinguish between the system's 'total' environment: everything not belonging to the system, and the system's 'relevant' environment. A system's relevant environment consists of:

> 'that set of entities outside the system, the state of which set is affected by the system or which affect the state of the system itself'.

This definition can easily be used in describing a system. With the criterion of interaction, the relevant environment is quickly reduced to manageable proportions. We must realise, however, that we shall then be dealing with a partial set out of the total environment, and that this involves certain implications for the research.

Some authors, including Ulrich [40], also use the concept of a super-system in analogy with the word subsystem. In studying a system, we can go into greater detail at a certain moment (raise the resolution level). We then study a subsystem of the system, and have to regard the rest of the system as the relevant environment. Or we can do it the other way round instead. We can lower the resolution level and change over from studying micro-phenomena to macro-phenomena. We then often move our system boundary and consider certain entities which first belonged to the environment as part of the system. We create a supersystem. We can then regard this again as a system, and so on. The supersystem of a system is, in fact, a set of entities which first belonged to the system, plus the entities which were at first part of the system's relevant environment.

Like the system, the environment may also have various characteristics. Since the environment's influence on a system may be of vital importance to the system, many authors have sought for the environmental characteristics.

In the case of organizations, Emery and Trist have characterised the properties of the environment by classifying environmental types [44]:

Type 1 is the least complex: the 'placid randomized environment'. This type has favourable and harmful influences on the system, divided relatively uniformly (with equal probabilities), unchangeable, and unchanging in their distribution. A typical example is the notion of the market in classical economics. As regards systems in such an environment there is no difference between strategy and tactics [45]: or, as Emery and Trist say 'the optimal strategy is just the simple tactic of attempting to do one's best on a purely local basis'.

Type 2 is more complex. It is still static, but the distribution of 'goods and bads' is now characterized by clusters: the 'placid clustered environment'. It corresponds to the market with the imperfect competition of economics. In this type of environment there is a difference between tactics and strategy. Since it contains clusters or accumulations, the system has a certain optimal or most favourable starting position: some positions are more attractive than others.

Type 3 is the 'disturbed reactive environment'. It is a type 2 environment, in which there is more than one organization of the same kind. The effects of other systems' actions on the system are taken into account in considering what course to adopt. This may lead to the practice of preventing other systems from reaching their goals in order to achieve one's own goal. This can be compared with the oligopolistic market of economics.

Type 4 is the most complex kind of environment, the 'turbulent fields'. It comprises all the elements of type 3, but is more dynamic. In type 3 the systems had dynamic properties, but the clusters in the environment were static. In type 4 these clusters change in their nature and/or their position; hence 'goods' become 'bads' and vice versa. Emery and Trist say 'the ground is in motion'.

3.5. REFERENCES

38. Miller, E. J. and A. K. Rice, *Systems of organizations*, London 1967.
39. Emery, F. E. and E. L. Trist, 'Socio-technical systems', in: F. E. Emery (ed.), *Systems thinking*, Harmondsworth 1969.
40. Ulrich, H., *Die Unternehmung als produktives sociales System*, St. Gallen 1968.
41. Allen, R. G. D., *Macro-economic theory*, New York 1968.
42. Lier, J. J. C. van, *Inleiding tot de thermodynamica*, syllabus Delft University of Technology, 1966.
43. Bertalanffy, L. von, 'General system theory', *General systems* I, 1956.
44. Emery, F. E. and E. L. Trist, 'The causal texture of organizational environments', in: F. E. Emery (ed.), *Systems thinking*, Harmondsworth 1969.
45. Schützenberger, M. D., 'A tentative classification of goal seeking behaviours', in: F. E. Emery (ed.), *Systems thinking*, Harmondsworth 1969.

CHAPTER 4

SYSTEM BEHAVIOUR

4.1. INTRODUCTION

In this chapter we shall deal with system behaviour and its characteristics [46]. When we speak of system behaviour we envisage the state of a system over a period of time. Concerning this definition, it should be realized that as regards the system's environment a change in the state of the system can only be observed by means of the change in its output.

Let us examine the concepts of the process, equilibrium, steady state, transient state, equifinality and stability. The concept of the process is difficult to fit into the framework of the concepts dealt with in this chapter. It seems to be more appropriate to system behaviour or the concept of state.

The concepts of equilibrium, steady state and transient state are properties of the state. They emphasize certain of its aspects so that there are advantages in differentiating between them.

Equifinality and stability are important characteristics of system behaviour. Systems possessing these behavioural characteristics are capable of perpetuating themselves. The recognition of these qualities is also important in system classification. Thus, a researcher studying a system of a certain class can try to make use of the general characteristics applying to this particular class. This will simplify his research. If, for example, a researcher observes stable behaviour in a system, he can avail himself of a general description for a stable system with respect to the system he is studying, specifying it according to the characteristics of the system in question.

4.2. PROCESS

The concept of the process easily causes confusion. This is due largely to the arbitrary use of the word 'process' in daily speech. It occurs, for instance, as: the production process, the thinking process, or the digestive process. When we speak of a production process, for example automobile manufacture, we are referring to a series of transformations effected with the various materials, finally resulting in a car. Miller and Rice [47] based their definition of a process on this idea. They say a process is:

> 'a transformation or a series of transformations brought about in the throughput of a system as a result of which the throughput is changed in position, shape, size, version or some other respect'.

In any case, this indicates the use of the concept of the process as the system's dynamic aspect. As Diesing [48] has said: 'Processes differ from systems in that they are composed of states rather than of elements and relations'.

On the assumption that behaviour is only observable from a system's output or acts, a process can also be defined as the system of actions of a system.

4.3. STATES

4.3.1. Transient state

This is the most general state we can envisage in a system. It is a state that changes in time (it is transient). It says little about the system's behaviour. Exceptions are systems where the change in state follows a predictable course. In general we tend to study systems in a transient state (for example, growth) because this is the most interesting. An example of a transient state is an alarm clock wound up by a spring. As the clock runs the spring relaxes and the state changes.

4.3.2. Steady state end equilibrium

Steady state means the unchanging state of an open system. This system has an input and an output, but the value of the corresponding state does not change. Man's body temperature is an example. It remains in a steady state because, even though there is continuous metabolism and heat is continuously lost to the environment, the temperature stays at 37° C.

A complete and all-embracing concept of state would consist of an infinite number of attributes of all elements for each system. It is inconceivable that all these would be in a steady state at the same time. Since we often disregard aspects, we can assume that a system exists for some time in a steady state with respect to a particular aspect. We can illustrate this again in man. As to his body temperature, there is a steady state. The size of his body, however, does not remain in a steady state during the period of his growth, since the states in which it exists are transient because of his growth. Bertalanffy [49] has an interesting translation for 'steady state' in German, 'Fliessgleichgewicht' (balance of flow). It is like a container holding a given volume of water, where the same amount goes out as comes in per unit of time. The weight of the container remains the same.

Equilibrium is the equivalent of a steady state. This term is used in dealing with a closed system. This means that a system is in equilibrium when the state does not change and there is no interaction with the environment.

4.4. EQUIFINALITY

This concept was introduced by Von Bertalanffy [50]. A system is said to have the property of equifinality if it can reach the same final state from different initial states and in different ways. For example, a given desired capital structure in an enterprise can be attained from various initial capital positions. There are different ways of doing this too: by increasing revenue at constant cost, by cutting costs with a constant revenue, and so on. Two comments are called for:

1. When open systems reach a steady state and show equifinality, the final state will be independent of the initial conditions.
2. The concept of equifinality does not apply to closed systems, because closed systems do not interact with their environment. The final state can only be determined by the initial conditions.

For some biologists, equifinality is a proof of vitalism (life). Von Bertalanffy remarks that this property can also be found in open, not entirely organic systems such as organizations.

4.5. STABILITY

Stability is a property found in many physical systems. One of the primary functions of the system is often the maintenance of stable conditions, since an unstable state is usually undesirable and frequently leads to catastrophes.

Examples abound: in biological systems illness usually indicates unstable behaviour. For instance, rapid variations in body termperature, blood pressure, or sugar content correspond with unstable conditions caused by some form of illness. The cobweb theorem in economics [51], representing an unstable state of supply and demand, is an example of an unstable system. Disturbance of price equilibrium by fluctuating supply in a market where the supply is

more elastic than the demand will make the price of a given product
fluctuate more and more (fig. 13).

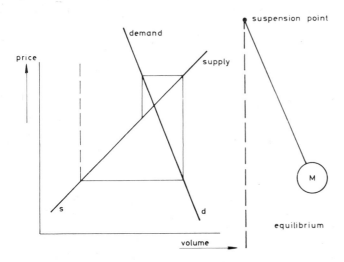

Figure 13. Cobweb.   *Figure 14.* A mechanical pendulum.

A mechanical pendulum is an example of a stable system in mechanics. If the pendulum is brought out of equilibrium, it will
ultimately return to equilibrium itself because of the retarding effect
of frictional forces at the point of suspension. Stability is a system's
ability to react in such a way that its response to external or internal disturbance disappears after the disturbance has been eliminated. A system can also be said to be stable when, after having
been brought out of its initial state by a disturbance, it returns to
its initial state after elimination of the disturbance [52].

Hall and Fagen [52] stress that a system can be stable in relation to
some aspects of the state, while at the same time it may be unstable
in other respects.

Forrester [53] describes a system as stable when it tends to return
to its initial condition after disturbance. He says it can return to its
initial state in different ways. The system can approach its initial

conditions gradually and reach them after some time. It can also overshoot its initial state and then approach it from the other side and, perhaps abated, start oscillating round the initial state. He describes the system as unstable if these oscillations become more pronounced. During such oscillations the system moves further and further away from its initial state.

In defining stability, there is thus always a question of returning to a given state. Some systems demonstrate this ability to a very great extent. They have a strong preference for a certain state, which may be called the desired state. Such stable, dynamic systems therefore possess the property of finality or teleology: of striving towards a goal (the desired state) from any initial state. Within the range of their stability they strive to attain this state.

The concept of stability includes more than merely returning to the initial state after disturbance. It is often impossible for the system to return to its initial state because conditions in its environment have changed. Yet the system can still behave stably by finding a state, different from the initial one, which is also stable.

Systems that behave in this way are called ultrastable [54]. They must possess the following properties:

1. They must be stable; they must be able to behave stably towards a number of disturbances in the environment, i.e. to return to their initial state after elimination of the disturbance.
2. If new kinds of disturbances emerge from the environment against which the system cannot maintain its existing behaviour, it makes a selection from among the available possibilities, by means of a selection process, of the behaviour that will lead to a new stability.

Since ultrastability need not appear only in living systems, we give an example of an artificial system: a component of the homeostat constructed by Ashby [55].

The starting point is a system interacting with its environment and displaying stable behaviour towards magnitudes essential to the system. This behaviour is created by continous interaction between the system and its environment. If the disturbance in the environment causes such an interaction that it is no longer possible for the system to behave stably, it can behave in alternative ways. In order to change its behaviour, the system has an instrument M, for measuring the changes in essential magnitudes in its environment.

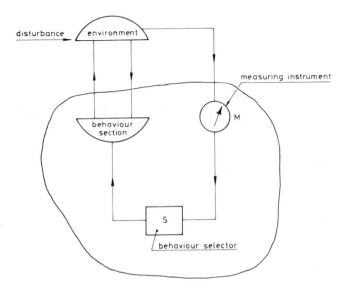

Figure 15. A subsystem of Ashby's homeostat.

In order to change its behaviour, the system has a selection mechanism S which, depending on signals from M, makes a choice from the available possibilities of behaviour (states), which change the interaction between the system and its environment. The experiment is as follows. When a disturbance is introduced in a system's environment, the value of the essential magnitudes changes. This disturbs the interaction between the system and its environment. First of all, the system will try to balance out the disturbance with the possibilities of behaviour available in the existing interaction between it and its environment. It will try to maintain its original state or to decrease the deviation from its original state. As soon as the total of the changes goes beyond the value which makes stable behaviour possible (this is measured by M), the system makes a random selection with S, of a new behaviour from the available possibilities.

This results in a new and different behaviour which also causes the system to interact differently with the environment. (The environment is already doing so as a result of the disturbance.) The system measures by means of M whether this new and changed behaviour creates a new condition of equilibrium in the interaction. If it does, it maintains the newly chosen state. If it does not, it will keep selecting another behaviour (another state) from among the available possibilities until a steady state is created in the interaction. If new changes occur in the environment, the system will keep on attempting to adjust itself to the changed conditions by selecting from the available states. By definition such a system is ultrastable.

We can go a step further with systems that consist of ultrastable subsystems. These are capable of partial adjustment, and are thus able to achieve or improve a desired interaction with the environ-

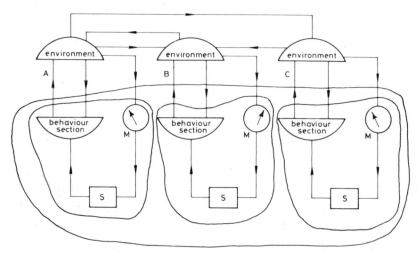

Figure 16. The multistable homeostat.

ment step by step. Such systems are called multistable. This is a behavioural characteristic of comparatively highly developed systems. In order to show that technical systems are also multistable, we will again base our example on the homeostat.

Three of these systems can be connected with one another, the interaction being effected by coupling the subsystems' environments. The starting point is again a stable state in the interaction between system and environment; in other words, in the three subsystems and their respective sub-environments. A disturbance is introduced into one of the sub-environments. Because of the interrelationship of the sub-environments, all sub-environments are affected. If the disturbance can be absorbed in the first instance by the interaction between subsystems and sub-environments, then the subsystems will not change their original state. If the disturbance crosses a certain limit in a certain sub-environment, its corresponding subsystem will change its behaviour analogously to the procedure described above. But because of this instability and because of the interrelationship, the interaction between the other subsystems and their corresponding sub-environments will be disturbed too. Reciprocal influencing and adaptation, however, will enable the entire system gradually to attain a new stable state via the interrelationships and with the help of the subsystems' ultrastable qualities. Systems with such properties are called multistable.

The concept of adaptation can be regarded as the highest form of stability (in the series stable, ultrastable and multistable), which can normally only be reached by self-organizing systems. This means that such a system is able to change its relations with its environment in order to retain an existing steady state, or else to find a new one.

4.6. REFERENCES

46. Katz, D. and R. L. Kahn, 'Common characteristics of open systems', in: F. E. Emery (ed.), *Systems thinking*, Harmondsworth 1969.
47. Miller, E. J. and A. K. Rice, *Systems of organizations*, London 1967.

48. Diesing, P., *Patterns of discovery in the social sciences*, Chicago 1971.
49. Bertalanffy, L. von, *Biophysik des Fliessgleichgewichts*, Brunswick, Vieweg 1953.
50. Bertalanffy, L. von, 'The theory of open systems in physics and biology', in: F. E. Emery (ed.), *Systems thinking*, Harmondsworth 1969.
51. Samuelson, P. A., *Economics*, New York 1967.
52. Hall, A. D. and R. E. Fagen, 'Definition of a system', *General Systems* I, 1956.
53. Forrester, J. W., *Industrial dynamics*, Boston 1961.
54. Klaus, A., *Wörterbuch der Kybernetik* I + II, 1965.
55. Ashby, W. R., *Design for a brain*, London 1954.

CHAPTER 5

INFORMATION AND ENTROPY

5.1. INTRODUCTION

Two concepts which appear frequently in the description of dynamic systems are information and entropy. In view of their nature these concepts are rather difficult to place in the context of the preceding chapters. A concept such as entropy can be regarded as a state variable; a definition for information is obviously harder to find.

We find these concepts difficult – hard to define uniformly and consistently. They concern the dynamics of a system. The processes with which they are associated largely determine the functioning of a system. In living systems they are often associated [56] [57] [58] with the origin of life and the continuation of existence. Our knowledge of these is still limited because of the complexity of these processes.

Of the two, the concept of entropy was introduced into science [59] first. It originated during the industrial revolution, in the days of the steam engine. At that time, the conversion of heat into mechanical energy was of great practical importance. It was essential to obtain maximum efficiency because of the limited energy sources. Considerable scientific attention was devoted to this subject. This led to the development of thermodynamics, a great contribution to which was made by Clausius and Kelvin. The second law of thermodynamics led to the introduction of the concept of entropy.

Entropy is an attribute of the state of the system under study and can be regarded as a measure of the differentiation in the system.

As we have already seen, in thermodynamics we speak of a closed system if there is no exchange of matter with the environment. From studies of isolated systems (systems which show neither matter nor energy exchange) Clausius concluded that in an isolated system all processes move toward a state of equilibrium, which is reached at maximum entropy. This conclusion has to be used with care as regards open systems because of their exchange with the environment. Later discussions will show that the term entropy must then be used differently.

Because of the difficulty of applying the concept of entropy to open systems, Stefan Boltzmann later produced a generalized formulation. He related it to the probability of the appearance of a certain state of a system. This formulation is far more general than that in thermodynamics. We speak of the most probable states, but the existence of less probable states, even if they last for only a particular period of time, is not ruled out. The formulation is more general because it is valid for all systems, whether open or closed, in many areas of science.

At the end of last century, scientists had many renowned discussions on the second law of thermodynamics and this concept of entropy. The physicist Maxwell devised a pencil and paper experiment rejecting the hypothesis in the second law of thermodynamics that the entropy of a system is always increasing. The experiment is called 'Maxwell's demon'. The demon stood at a door in a wall separating two compartments of a barrel filled with gas. The idea was that by opening and closing the door in a special way, he would get the fast molecules of the gas on one side and the slow ones on the other side. This would decrease the entropy of the system instead of increasing it. Up to 1929 no one could disprove this 'experiment', though in reality it could not be carried out. Then Szilard [60] showed that it did not produce any logical conclusion. Instead, he introduced the concept of information. He claimed that the demon would have to know the velocity of the approaching molecules in order to open the door just at the right moment. In the conditions in which the demon had to work, this would be impossible. So the demon could never do his work according to plan, and the hypothesis of the second law remained true. This made Szilard the first person to establish a relationship between entropy and information.

The concept of information was not elaborated further or used scientifically until after 1948. Especially after its formalization and quantification by Shannon [61] in 1948, the development of information theory went ahead more rapidly. Shannon arrived at a mathematical definition of the concept of information. Since then many writers have studied relations between entropy and information, stimulated by the observation that the mathematical model developed by Boltzmann for entropy and that developed by Shannon for information closely resemble each other. Other sciences, especially cybernetics (Wiener describes cybernetics as the science of information), have benefited from this development.

5.2. INFORMATION [62]

5.2.1. Communication [63], [64], [65]

Communication, and hence information, is a commonplace thing. We use it in a wide variety of ways: talking, reading, looking. It not only embraces the use of language (verbal communication), but also gestures or actions (signs) which tell us something (nonverbal communication). By communicating we try to transmit something: feelings, orders, instructions, and so on. This transmission, and hence the communication process, can be represented as a model (fig. 17).

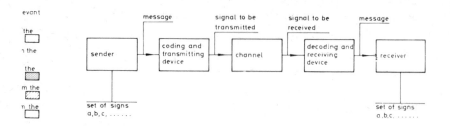

Figure 17. The communication process.

A person (or a machine) plans, for example, to address another person (or machine). For this purpose he (the sender) has at his disposal a number (set) of signs with which he can describe concepts and events taken from reality. The sender now chooses from this set a certain arrangement which, in his view, has meaning and seems adequate for addressing the other. This arrangement of signs is the message.

In order for this message to reach the receiver, it has to be converted into a transportable form. This is done by the coding and transmitting device. It first codes or converts the message according to a set of rules (the code) into a physically transportable form. In the case of one person addressing another the code is the language that is used, containing a set of rules with which our thoughts (set of signs) are converted into spoken language (for the moment disregarding non-verbal communication). The vocal cords form the transmitting device: they convert the coded message into a physically transportable signal (sound vibration). The coded message converted into a physically transportable form is known as a signal.

The signal is transported along the channel to the receiver's decoding and receiving device. The channel is the physical transportation route, for instance electric cable or air. The channel has limitations on its transmitting capacity, which is finite for each channel. If the sender, or, in fact the transmitting device, produces more signals than the channel can cope with, the receiving device receives only part of the signals, corresponding to the channel's maximal capacity.

As the channel's properties are often not ideal, there is also a risk of the signal being affected by interference in transmission which may distort the specific contents of the signal. The signal received is then converted into a message by the receiver's decoding and receiving device. The signal is first received (by the outer ear, the ear drum, etc.) and then decoded with the same code used by the transmitter. The signal is then converted into a message. This message consists of a combination of signs, but this time of signs from the receiver's set. If both sets are equivalent and transmission of the signal along the channel has not undergone too much interference, the message received will be the same as that sent, and a communication process will thus have taken place. The message as received then means the same to the receiver as to the communicator. The message 'Could you give me a light' has been received as such by the receiver and he has understood it.

The code is of the greatest importance to a communication process. In fact, a code is a range already containing all conceivable messages in advance. For example, someone knowing the code of a card game knows that the front of the card, which he sees only from the back, contains only one of 52 possible images. In theory, all possibilities of written communication are already intrinsically contained in the alphabet. This indicates that the supply of signs is another limiting factor. With the available combination of alphabetical signs, a word can be formed that is in the vocabulary of the language (for example: entelechy) but is not in the receiver's set. He needs a dictionary to put it into a context understandable to him: 'the characteristic of matter to strive for self-realization in its form'.

5.2.2. *Syntax, semantics, pragmatics*

Communication is the transmission of signs, and as such is a part of semiotics (general theory of signs). We distinguish three areas in semiotics:

1. syntax, the area of signs and combinations of signs;
2. semantics, the area of referring to a world, to reality;
3. pragmatics, the area of the use and effect of signs.

Syntax is concerned with the formal theory of signs, determining signs and the rules for combining them (vocubulary and grammar). The important factor is not content and meaning, but the correctness of signs and their combinations. Semantics deals with the meaning and content of signs, the influence of content on meaning and on the presence or absence of content. In pragmatics we study the effect of using signs on the receiver's behaviour and their effectiveness relative to the desired result.

With the aid of this tripartite division, let us take another look at the communication process described in the last section. We have already seen that this is an idealised representation of a communication process and that in reality there are many distorting or interfering influences. In order to classify these distortions (noise) we use the tripartite division. This classification makes it easier to find adequate measures to deal with them.

Extraneous influences on the channel or the coding or decoding device, which change the sequence or combination of signs are known as syntactical disturbances. They are generally of a technical nature. They can be remedied by technical means or by introducing redundancy (see 5.2.4).

Semantic disturbances are a second category. They emerge when the sender's and receiver's sets of signs are dissimilar (for instance the sender uses English to contact a Chinese receiver who does not understand English). The message will probably be technically correct when it reaches the receiver but will be incomprehensible to him. The communication process has either not taken place or is incomplete.

The third category is that of pragmatic disturbances [66]. The sender must be regarded as having a given expectation of the effect his message will have on the receiver, and his reaction to it. If the receiver does not respond accordingly, there is a pragmatic distortion of the communication process.

5.2.3. Signal, message, information

We now discuss the above concepts in greater detail as regards their important role in the communication process. A signal is a message converted into a physically transmissible form (for example electric current along a telephone wire). The message is a combination of signs built up by certain rules (syntax) by the sender and referring to a reality existing in the sender's thoughts (semantics). The sender wishes to describe this reality by means of his message and transmit it to the receiver. He wants to evoke a certain response by the receiver (pragmatics). The concept of information is now coupled to the concept of message. The question is, when does a message contain information for the receiver?, i.e. when does the receiver do something with the message?

This happens if the message removes the receiver's uncertainty in some way or other. The receiver may, for instance, hear that his mother is ill, or he may be given certain instructions. Obviously, not every message will eliminate all uncertainty. If, for instance, instructions are received, questions still remain such as: how, when and where? The receiver will then ask for further information or go ahead on his own accord, and in the latter event pragmatic disturbances are soon likely to occur.

Similarly to communication, we can speak of syntactic, semantic and pragmatic information. Syntactic information is information without meaning, content or effect for its receiver. It is merely a combination of signs, arranged in accordance with grammar and vocabulary. Mathematical and statistical information theory (see 5.2.5) deals particularly with syntactic information.

We refer to semantic information when the message also has a meaning. It then refers to a reality. An example will clarify the difference between syntactic and semantic information [62]: Compare the following two messages: 'the enemy launched an attack with three batallions at five in the morning' and 'the enemy launched an attack with several batallions in the morning before noon'. The first sentence contains more semantic information than the second. But the latter has more letters and words and hence contains more syntactical information than the former.

The concept of pragmatic information is the most embracing. We deal not only with a signal, the series of signs and their meaning, but also with the effect the message has on its receiver. It is the kind of information which, with equal messages, i.e. equal signals and content of signals, can make two different systems, two different receivers, react in totally different ways. This therefore means that we also include the state, which is a result of a variety of prior systems inputs, in studying the information concept.

To sum up, we can define communication, like Brönimann [65], as follows:

1. Communication is the transmission of messages between entities or subsystems of a system, or between systems.
2. Communication means not only the technical transmission of signals and messages, but also the receiver's reaction. Like information we can investigate communication from syntactic, semantic and pragmatic points of view.
3. We can envisage communication as from sender to receiver. Also envisage in both ways: from sender to receiver and from receiver to sender. Information is the object of the communication.
4. A message contains information when it removes uncertainty in some way.

5.2.4. Redundant, relevant and irrelevant information

These concepts arise when someone, let us say the researcher, is faced with a problem. To solve this problem, he can define the quantity of relevant information. This is the minimum amount of information he needs to solve the problem properly. In order to obtain this information, he will seek data, make inquiries, ask questions and so on, thereby initiating a communication process.

Thus the researcher starts obtaining information. The process can be presented as follows: he asks a question. Thereupon he receives a message containing a quantity of information. This information may be partly irrelevant to the problem. He now asks a second question (perhaps triggered off by the first one; this involves a learning process), and receives a second message and hence a second amount of information. Here again, part will be relevant, part irrelevant. Part of the relevant information in the second message may, however, already have been contained in the first message. Although our definition so far indicates that this part of the information at this point is, strictly speaking, no longer of any use to the researcher – since it does not remove any uncertainty – we nevertheless call it information. This is redundant information. Put briefly, irrelevant information is information which the researcher does not need, and redundant information is that he already has.

This is shown in figure 18 [67].

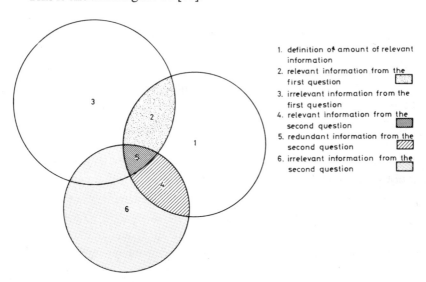

1. definition of amount of relevant information
2. relevant information from the first question
3. irrelevant information from the first question
4. relevant information from the second question
5. redundant information from the second question
6. irrelevant information from the second question

Figure 18. Redundant, relevant and irrelevant information.

Redundancy might appear to be undesirable, but this is only partially true. It would be undesirable if there were no syntactical or semantic distortions. The lack of redundancy means that a mistake in the message, and hence in the amount of information, immediately has adverse effects. The message does not reach the receiver correctly; it may even be unintelligible and as a rule the receiver will give the wrong response to the message.

If a message is purposely made redundant the receiver may have the opportunity to verify it. The message already contains information known to its receiver. Comparing it with the known information provides an initial check. This method is used especially for technical purposes (for example in the case of computers working with a check-digit). There are various ways of doing this. An example of syntactic redundancy is: tthhiiss iiss aa lliinnee. Repetition of each individual letter has, in fact, doubled the syntactical information. This method is widely used in telegraphy to suppress noise in the cable. It provides the telegraphist with a check. Redundancy can also be introduced semantically. In the sentence 'this is a straight line forming the shortest connection between two points', the phrase 'forming the shortest connection between two points' is, in fact, redundant information. If for some reason someone does not understand what is meant by 'straight line', this is clarified by adding 'forming the .....'. Redundancy is generally a means of safeguarding a message against semantic or syntactic distortions: it is used to reduce the effect of interference in the communication process.

Human language is also very redundant. For example, the following test was made [68]: In the preparations for an examination, the students were randomly divided into four groups. Before the test, eight papers had to be studied. Four versions of the papers were handed out. The first one was the original paper. The second was two-thirds the length of the originals. It was reduced by leaving out articles and other superfluous words. The third was reduced to one-third of the original length, while the fourth was an abstract written by experts. There were no significant differences in the results of the students who studied the first three versions. Students who had studied the abstract scored significantly lower in the examination on this paper. The conclusion is that two-thirds of our use of language is redundant. But this redundancy seems to be essential for enjoyable, understandable use of language.

In this context, Van Peursen [62] deals with what he calls 'the psychological information range of information-redundancy ratios'. It is: 'confusion - overload - surprise - news - not expected - recognizable - familiar - superfluous - known-dull - repetition - unnecessary'. In this way he wishes to indicate the possibility of determining the degree of desired redundancy. He postulates that too much redundancy makes reality tedious, while too little results in over-exertion, confusion, and an overload of information.

5.2.5. *Mathematical information theory* [61]

So far our discussion of the information concept has mainly been qualitative. As a result, a formal comparison between messages concerning information content, efficiency of transmission and so on is difficult. In order to make this comparison possible, we have to formalize the concept of information and quantify the amount of information. The mathematical information theory tries to do this.

The first step is to develop a unit, a measure, for the amount of information, and the term 'bit' (a contraction of binary digit) is introduced for this purpose. The bit is in general use as a unit in mathematical information theory. Physically, it is dimensionless.

The 'bit' is defined as follows. We assume that the receiver knows the total amount of facts or messages he could receive, i.e. he knows all the possibilities. Let us give an example. Take a stack of 32 cards. The receiver knows all the cards and has to solve the following problems: 'find in the quickest possible way any card specified by the sender, only asking questions that can be answered by 'yes' or 'no'. The receiver can find the answer by asking no more than five questions, viz: 'Is the card one of the first 16: answer 'yes' or 'no', and so on. By constantly dividing by two he will find the card in the quickest possible way. The receiver has now received 1 bit of information per question, or 5 bits of information in total. The number of bits of information embodied in something is in fact the number of yes/no questions that have to be asked in order to find the required possibility.

In this case, when the probability of the same card occurring is the same in all 32 cases, the following relation can be deduced for determining the informational value:

$$I = K \cdot {}^2\log N \text{ (bits)}$$

in which:

I = informational value
K = constant (usually $+1$)
N = the total number of a priori equally probable possibilities.

Hence in our example:

$$I = {}^2\log 32 = {}^2\log 2^5 = 5 \text{ (bits)}$$

This has introduced the 'bit' with a rather specific example. No differentiation was made between the possibilities that exist. They all had the same probablility of appearance ($p_i = 1/N$). We can, however, also generalize the quantity of information. We start with a specified number of possibilities N, and the probability of each of these possibilities occurring p_i, ($i = 1, ..., N$). If, in this situation, the answer (if existing) to a question is given for which the probability of occurrence is exactly 0.5, then the receiver will obtain one bit of information in this answer.

In order to clarify this, let us give two examples:

1. Four possible answers: a, b, c, d,
   the probability of occurrence is 0.5; 0.2; 0.2 and 0.1 respectively;

the quantity of information comprised in answer a is then:

$$I_a = -^2\log p_a = -^2\log 0.5 = 1 \text{ bit, and so on.}$$

We can also define the average information for each possibility. It is found by adding the amount of information for each possibility multiplied by the relative weight (this is the probability of occurrence):

$$I_{ave} = \Sigma P_i \cdot I_i$$

In this example this becomes:

$$I_{ave} = -0.5^2 \log 0.5 - 0.2^2 \log 0.2 - 0.2^2 \log 0.2$$
$$- 0.1^2 \log 0.1 = 1.76 \text{ bit}$$

With these possibilities to choose from, one would then receive on average 1.76 bits of information per question.

2. If a young man asks a girl to marry him, the amount of information is usually much less than 1 bit. This is because the young man would not usually ask this question until he was fairly sure of the answer. If the probability of the girl saying 'yes' is 0.95, the amount of information transmitted to the man by the girl's 'yes' is the logarithm to the base 2 of 0.95:

$$-^2\log 0.95 = 0.074 \text{ bit}$$

In the sections dealing with transmission of information we observe that transmission never causes a loss of information to the sender. The sender's amount of information is not reduced after transmission (he still knows what he has transmitted), while the receiver's has been increased by the information transmitted. Information transmission does not have the same characteristics as, say, energy transmission, in which the sender's state has changed after transmission.

5.3. ENTROPY [62]

5.3.1. Thermodynamic entropy concept

As stated above (5.1), the concept of entropy was introduced in thermodynamics (the physical theory of heat movements). Thermodynamics has two laws: the first relates to the conservation of energy; the second states that in energy processes entropy always increases.

The law of the conservation of energy states that the total amount of energy in a system does not change, but says nothing about how this energy is distributed among its various forms, such as heat, mechanical energy, electrical energy and pneumatic energy.

Experience has shown that energy is consumed; mechanical energy, for instance, disappears after some time and is converted completely into heat which is no longer usable (otherwise it would be possible to develop perpetual motion (Latin: 'perpetuum mobile'). These findings are generalized in the second law of thermodynamics. Clausius gives the following formulation: heat can never pass from a low temperature state to a higher temperature state by itself. In this formulation two magnitudes stand out: an amount of heat (Q) and the temperature of the matter that contained this heat (T). We can now deduce from the second law that the quotient of Q and T: Q/T, in passing between different states, must always increase. This quotient is the 'reduced heat' or entropy. Hence entropy always increases.

What, then, is this concept of entropy [57]? Firstly, it is not a theoretical concept but a measurable magnitude, such as the length of a bar or the temperature of a body. At absolute zero (0 Kelvin or -273.16 °C), we state that the entropy of each system is zero. If we subsequently bring the system into another state by supplying energy in slow, reversible[1] stages (even if the system thereby changes

---

1. Supplying energy in reversible stages means that we return to the same initial state if we take the same amount of energy away agian.

its physical or chemical composition or separates into two or more parts with different physical or chemical compositions), the entropy increases by an amount calculated by dividing each small amount of heat supplied by the absolute temperature (in degrees Kelvin) at which it was supplied, and by adding all the small amounts together. For example, if we melt solid matter, the entropy of the system increases with the heat of fusion divided by the (absolute) temperature at the melting point. Hence the unit of entropy is the Joule/Kelvin (J/K).

This can be illustrated with an example from thermodynamics [56]. Let us take a system consisting of two interconnected bodies. The first has a temperature T_1, the second T_2 ($T_1 > T_2$). Since T_1 is higher than T_2, a current of heat will pass from body 1 to body 2, until both reach the same end-temperature T_3 ($T_2 > T_3 > T_1$).

The heat content of the system has not been changed by this heat transport (first law).

Figure 19. Heat current Q from a body with a higher temperature T_1 to a body with a temperature T_2.

In order to present the process comprehensibly, we now quantify the heat-transfer process. At first, a small quantity of heat dQ at temperature T_1 is transferred to body 2. This will hardly change the temperature of 2 (T_2). However, the entropy has increased because dQ/T_2 is greater than dQ/T_1 since $T_1 > T_2$. We can continue in this way until the equilibrium temperature T_e is reached. Since heat passes continuously from a higher temperature to a lower temperature, the entropy goes on increasing by the difference:

$$dQ/T_1 - dQ/T_2 \quad (J/K)$$

Finally, the system will reach the equilibrium temperature T_e ($T_1 > T_e > T_2$) with both bodies at this temperature. The total increase in entropy can be calculated by totalling the differences calculated above. This addition will acquire the character of an integral. The total increase in entropy ΔS is:

$$\Delta S = \int_{T_1}^{T_e} dQ/T - \int_{T_2}^{T_e} dQ/T = \int_{T_1}^{T_e} dQ/T + \int_{T_e}^{T_2} dQ/T = \int_{T_1}^{T_2} dQ/T \ (J/K)$$

It follows from the above reasoning that in an isolated system[2] (where there is no exchange of matter or energy) all processes move toward a state of equilibrium, which will be reached as soon as the entropy in this system is maximal.

A well-known statement in this context was made by Clausius [58]. He regarded the universe as a closed system and said 'Die Entropie der Welt strebt einem Maximum an' (the entropy of the Universe tends towards a maximum state). This statement has given rise to speculations about the 'death by entropy' of the universe; i.e. haeuniverse developing so that a general levelling occurs, and finally all conditions will become equal everywhere. Life can then no longer exist. The question is: is the universe in fact closed?

5.3.2. Statistical entropy concept

The greatest objection to the entropy concept introduced above is that its validity is limited to isolated systems. To make it applicable to systems with fewer limitations, it has to be generalised.

This generalization was worked out by the physicist Boltzmann. He already noticed about a hundred years ago that two macroscopically similar states between which no difference can be found with normal

---

2. In thermodynamics, a differentiation is made within the concept of a closed system (see 3.2) between a 'closed system', i.e. one in which there is no exchange of matter with the environment (see 3.2) and an 'isolated system' where, in addition, no energy is exchanged with the environment.

measuring instruments may be very different microscopically. Macroscopically, two systems or two states may resemble each other like two identical drops of water. Microscopically, the position and motion of every molecule will differ from moment to moment and from drop to drop. Continuing this parallel of drops of water, we can ascribe a state to each water molecule in the drops (the micro-state). If we call the number of possible, equally probable micro-states g, then Boltzmann inferred that the entropy of the drop is proportionate to the natural logarithm of g.

This gives the equation:

$$S = k \cdot \ln g$$
$k = $ Boltzmann's constant

The increase in entropy implies an increase in g, i.e. an increase in the number of equally probable micro-states.

We can also define the entropy concept in non-equally probable micro-states. The average entropy (see also average information: 5.2.5.) will then be:

$$S = k \cdot p_i \cdot \ln p_i$$
in which: p_i is the probability of occurence of micro-state i.

The second law can now be phrased as follows: a system strives to reach the most probable state. In this definition the concept of heat no longer appears. The thermodynamic definition can be reduced to the above definition and relates, in fact, to a special case.

This is not absolute formulation, as is expressed by 'strives'. We speak of the most probable state, but the temporary existence of less probable states is not precluded.

In general terms, entropy can be understood as a measure of the disorganization in a system. By definition, we describe a state as more disorganised if it can be achieved in more ways. The second law of thermodynamics expresses the tendency for the system to strive towards a state of maximum probability: maximum disorganization. This striving for maximum disorganization may be counteracted by the striving towards a minimum energy content in the system corresponding to the striving for organisation. In thermodynamics this is expressed by the concept of 'free energy'.

5.4. RELATION BETWEEN INFORMATION AND ENTROPY [62]

It is perhaps not yet completely clear to everyone that a relation can be established between information and entropy. Szilard's observations (5.1.) about Maxwell's demon, however, already suggest this relation. We discuss it because present-day literature on organizations, again and again, whether justified or not, contains arguments concerning the analogy between entropy-energy processes in thermodynamics and disorganizing processes in organizations (entropy as against information). Wiener says: 'Just as the amount of information in a system is a measure of its degree of organization, so the entropy of a system is a measure of its degree of disorganization: and the one is simply the negative of the other.' It would seem useful to emphasise the background to these arguments. In our opinion, the greatest care should be taken in formulating concepts about such processes.

Entropy is a measure of the degree of disorganization in a system. A system's tendency to strive toward a state of maximum disorganization (maximum entropy) is expressed by generalization of the second law of thermodynamics. The contrary striving for organization is shown by the endeavour to collect information. Information can be considered as a measure of organization in a system. If only disorganizing processes (an increase in entropy) influenced a system it would quickly attain a state of maximum entropy.

Collecting information makes it possible to 'create order in chaos'. Open systems around us present this pattern. Open systems are therefore capable of gathering information in order to restrain the striving for disorganization. In a certain sense, entropy and information are each other's opposites. Many biologists have tried in this way to explain why an organism can stay alive in spite of hostile conditions in its environment.

This reasoning is enforced if we compare the mathematical models developed in respect of information (5.2.5.):

$$I_{ave} = K \cdot \Sigma p_i \cdot {}^2\mathrm{log} p_i$$

p_i is the probability of the occurrence of the i-th possibility.

And for entropy (5.3.2.):

$$S_{ave} = -K \cdot \Sigma p_i \cdot \mathrm{ln} p_i$$

p_i is the probability of occurence of micro-state i.

Structurally these models are identical, i.e. apart from a constant factor they can be carried over into each other. They differ dimensionally, though some scientists contend that the differences are surmountable. The agreement between these two models strengthens the opinion that the concepts of entropy and information have a dual character. Van Peursen [62] says: 'in fact entropy and information are two measures of the same phenomenon; they are two sides of a coin'. Entropy is a measure of arbitrariness, disorganization, the unknown, while information can be regarded as a measure of the known, organized and specific. Many writers therefore say that if an (open) system is left to itself, it will strive towards the most probable state, i.e. maximum entropy. This process can be counteracted by discarding some of the entropy to the environment, by taking in negative entropy, i.e. by acquiring information.

This reasoning is often used with respect to social organizations considered as systems. Katz and Kahn [69] state that if these systems did not import 'energy'[3], they would degenerate and strive towards the most probable state, which means 'death by entropy'.

3. 'energy' in this sense has the character of information: information can be obtained by means of energy.

The living systems already mentioned are also proof of this hypothesis. During their existence they are capable of attaining ever
higher forms of differentiation through the continuous introduction
of information (cf. the development of man; from baby to adult).
At a certain moment a stable level is reached which is then followed
after a certain time by sudden collapse (death). A minor defect in
the 'order' is enough for this to happen.

Yet, in using such reasoning by analogy, some marginal comments
are called for. The most important is that thermodynamic concepts
only apply to thermodynamic systems. In principle these are closed
(in relation to the transfer of matter). As against this, organizations
are completely open systems. The very existence of an environment
and interaction of the organization with this environment is the
most essential aspect for continuance of the organization. Great
caution is therefore needed in applying these concepts of information and entropy to open systems.

As regards their application it might be mentioned that so far we
know of no particular operational applications. They often go no
further than rather vague, qualitative analogical arguments at best
clarifying to some extent the direct concept but operationally un-
fruitful. Questions such as: 'what is the information or entropy content of an organization?' have not yet been answered satisfactorily
in our view.

5.5. REFERENCES

56. Schrödinger, E., 'Order, disorder and entropy', ch. 17, in: W. Buckley
(ed.), *Modern systems research for the behavioral scientist*, Chicago 1968.
57. Brillouin, L., 'Life, thermodynamics and cybernetics', ch. 18, in: W.
Buckley (ed.), *Modern systems research for the behavioral scientists*,
Chicago 1968.
58. Raymond, R. C., 'Communication, entropy and life', ch. 19, in: W.
Buckley (ed.), *Modern systems research for the behavioral scientist*,
Chicago 1968.

59. Boiten, R. G., 'Cybernetica en samenleving', in: *Arbeid op de tweesprong*, The Hague 1965.
60. Szilard, L., 'Über die Entropieverminderung in einem thermodynamischen System bei Eingriffen intelligenter Wesen', *Z. für Physik*, 1929, p. 840.
61. Shannon, C. E., 'The mathematical theory of communication', *Bell System Technical Journal*, 27, 1948.
62. Peursen, C. A. van, C. P. Bertels and D. Nauta, *Informatie*, Utrecht 1968.
63. Mirow, H. M., *Kybernetik*, Wiesbaden 1969.
64. Coenenberg, A. C., *Die Kommunikation in der Unternehmung*, Wiesbaden 1966.
65. Brönimann, C., *Aufbau und Beurteilung des Kommunikationssystems von Unternehmungen*, Berne, Stuttgart 1970.
66. Watzlawick, P., J. H. Beavin and D. D. Jackson, *Pragmatics of human communications*, New York 1967.
67. Peters, J., *Einführung in die allgemeine Informationstheorie*, Berlin 1967.
68. Cherry, C., *On human communication*, 2nd ed., London 1961.
69. Katz, D. and R. L. Kahn, 'Common characteristics of open systems', in: F. E. Emery (ed.), *Systems thinking*, Harmondsworth 1969.

CHAPTER 6

MODELS

6.1. INTRODUCTION

In the introduction (1.1.) we pointed out that at present there is a great deal of interest in applying a systems approach for solving social, political and environmental problems. In order to describe these large-scale, complex, interactive systems, the systems approach makes use of models. Symbolical representations are used to describe these complex systems, so that conclusions regarding the effects of alternative system configurations can be drawn quickly and effectively. The process of constructing, of making models themselves, is also being more and more understood as a direct extension of scientific method [70].

It took a long time for the concept of the model to be scientifically worth while. The use of models in science was long looked down upon. Only after the formalization of the concept of the model, and the introduction of the concept of isomorphism, was the model integrated into scientific methods. Since then the concept of the model has been of great scientific value because of its many functions [71].

The model has played (and still plays) its most important role in physics. Without it, the explanation of magnetic and electrical phenomena (in the 19th century) would have been more difficult. Ether is a well-known example of a clear physical model used for clarifying the phenomena of light.

This model, however, was superseded by Maxwell's development of the electro-magnetic theory. The empirical model (using the ether as the medium for propagating electro-magnetic waves) was replaced by a formal model (Maxwell's). Graphical clarity was replaced by abstraction.

This was once again emphasized by Poincaré, who proved that the existence of a single mechanical model for a physical process implies the existence of an infinite number of such models. This also made it clear that the model should be regarded as an artificial thing, constructed by the researcher, instead of a natural product. The model is regarded more and more in an operational sense. It is interesting only in so far as it is appropriate [72].

As a result of the explicit use of models in science, systems thinking developed, with its integral approach to complex problems. The use of models made it possible to approach these problems more effectively in general terms.

6.2. THE MODEL

What exactly do we mean by a model, by the use of models? The essence of using models is that a material or formal image of a system is made which is easier to study than the system itself. This image is then used as a model of the system. The model must then obviously contain information about the system. Hence there must be a certain resemblance between the model and the system.

We can construct models for various purposes [73]:

1. As regards certain phenomena no theory exists as yet. In studying these phenomena we can seek a science in which a theory exists and which has major characteristics in common with the area of research in question. We can then use a model to develop our knowledge. This is what happens in neurology, where it has been observed that the functioning of the central nervous system bears resemblances to the functioning of an electronic computer. By

studying a computer in operation, an effort is made to reveal the properties of the central nervous system.

2. As regards certain facts, we have a completely valid theory, which cannot be solved mathematically with our present mathematical techniques. The fundamental concepts are then interpreted in the form of a model using simplified assumptions, so that the equations can be solved. The theory of harmonic oscillators for studying heat conduction is an example.

3. We can use a model for relating two unrelated theories, by using one as a model of the other, or by introducing a common model relating them together.

4. Some theories are properly confirmed, but are incomplete as regards the class of phenomena they explain. In order to ensure greater completeness, we can set up a model. Studying this model often makes it possible to achieve completeness.

5. If new information is acquired regarding a given phenomenon, we also construct a model in order to convince ourselves that the new and more general theory also embraces the old theory. With this model we then show that the old theory was a specific case of the new one.

6. A theory concerning certain facts need not explain these facts. Models can provide the explanation (for example the wave and quantum presentation of light, or statistical mechanics).

7. We sometimes want to study objects too small, too large, too far away, or too dangerous for experiments with them. We then experiment with models of such systems. Such models have to be representative enough to obtain the desired information.

8. In order to have a formalized theory available, it is often useful to use a model that visualizes or realizes this theory.

9. The theoretical level is often far removed from the level of perception; ideas cannot be translated directly into observations. Models are then introduced to bridge the gap between these two levels.

Apostel [73] says that models are introduced as a function between theory and theory, experiment and theory, experiment and experiment, and between thought structures and those using these structures. In all cases this has been done to produce new results, or to verify results with the aid of earlier findings, or to demonstrate relationships.

Let us now try to distil from the foregoing some general characteristics of the concept of the model, which might lead to a generalized definition. In using models, there are always at least two systems: that being studied S, and its model M. In principle each of these two systems, M and S, is independent.

This is epistemological independence; i.e. what is known about M is not information ultimately obtainable from S. This implies that there must be no interaction, either direct or indirect, between M and S. Someone uses M as a model of S in order to obtain information about S. All this gives the definition of a model [72]:

'If a system M, independent of a system S, is used to obtain information about system S, we say that M is a model of S.'

6.3. ISOMORPHISM AND HOMOMORPHISM [74]

In order for adequate information to be obtained about the unfamiliar system S, model M has to fulfil certain requirements. These can be summarized as: M must resemble S in its structure. Generally speaking, this means that the model and the system are isomorphic (from the Greek: having the same form). Since we want to obtain information about the original system through the model, and this is best done by using an isomorphic model of the system, isomorphism is said to be the essence of the concept of the model.

Two isomorphic systems have corresponding structures. Hence, the number of relations and the way they are arranged are the same. The pattern of the relations must be the same. This correspondence can be formalized with the concept of mapping from set theory. A mapping is a special function transforming one system into the other. This can be clarified with an example (fig. 20).

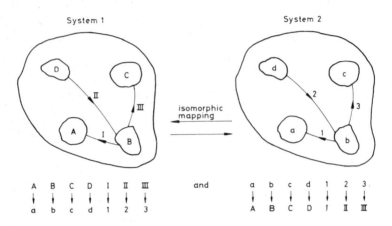

Figure 20. Isomorphic mapping of system 1 on system 2 and vice versa.

The entities and relations of system 1 are mapped onto those of system 2. This mapping is also possible the other way round, the structure always remaining unchanged. The mapping is 'one-to-one and onto', because each entity of system 1 has assigned to it an entity of system 2, and the relations are mapped accordingly (thus relation I between A and B is mapped on relation 1 between a and b). We speak of mapping as being both symmetrical and binary [72]. This means that the mapping can be made in either direction, and that two elements are always involved. We can now define isomorphism as follows:

> 'two systems are isomorphic when there is a one-to-one and onto mapping transforming one into the other with the con-servation of the relations'.

The concept of homomorphism can easily be clarified if we regard isomorphism as a special case of homomorphism. Homomorphism is a generalization of isomorphism. In homomorphism there is a 'many-to-one' mapping, and in isomorphism a 'one-to-one' mapping (fig. 21).

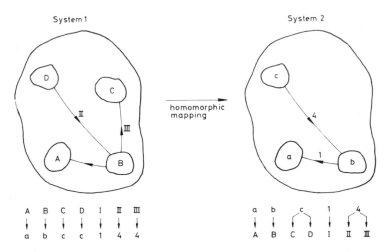

Figure 21. Homomorphic mapping of system 1 on system 2.

The homomorphic mapping of system 1 on system 2 has significance in one direction only. The other way round, problems arise because it is not known, for instance, where c should be mapped, on C or on D? Hence the homomorphic mapping is many-to-one. It is binary but not symmetrical.

It should be pointed out that in the case of two homomorphic systems a number of relations are omitted, as a simplification. Since we are using models which should preferably be easier to study than the system itself, we will generally use models which are homomorphic with the system under observation. In practice, however, the concept of isomorphism is generally used (perhaps wrongly so). One reason is that in studying phenomena, models of an aspect system are usually made.

The researcher tries to make an isomorphic model of this aspect system. But as the aspect system itself is a homomorphic model compared with the system, the model produced by the researcher is homomorphic with respect to the system in question.

A well-known example of isomorphic (homomorphic) systems is the 'electrical-mechanical analog' [75]. The mechanical system (fig. 22) consists of a revolving mass m, attached to a shaft to which a frictional force can be applied and which is attached to another shaft by a spring. One turns the latter without the mass m slightly, after which the shaft is released. Since this has loaded the spring to some extent, the shaft without the mass m will begin to turn. The spring passes this rotation onto the shaft with the mass m, which starts turning as well. The rotation of the mass m as a result of rotation of the shaft can be described by a second order differential equation of the form:

$$a\frac{d^2y}{dx^2} + b\frac{dy}{dx} + cy = f(x)$$

in which a, b and c are defined respectively by the mass of the body m, the frictional force F on the mass and the tension/constant k of the spring.

The electrical system (fig. 23) consists of an inductor, a capacitor and a resistor, connected in series. It now appears that the variations in amperage in this circuit, after initial disturbance by a voltage source, can be described by the same kind of second order differential equation as the mechanical system. a, b and c are then defined respectively by the induction L of the inductor, the resistance R and the capacity C of the capacitor. If we choose the values of the parameters of the mechanical system: m, F and k, corresponding to those of the electrical system, L, R and C, then the variations in the movement of the mass correspond to the variations in amperage in the circuit.

We are now dealing with 3 isomorphic (homomorphic) systems: the mechanical system, the electrical system and the formal system (differential equation) which, because of isomorphism, can all be

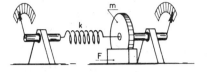

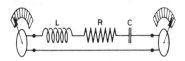

Figure 22. Mechanical system.   *Figure 23.* Electrical system.

considered as models of one another. This presents considerable advantages. For example, if a mechanical system is not appropriate to a particular study, then the formal system or the electrical system can be used to study it. This is, of course, also possible the other way around. In this way, the easiest means of solving the problem can be selected.

6.4. MODEL CONSTRUCTION

The model construction process is summarized in the model cycle [76]. This visualizes the three phases of model construction: abstraction, deduction and realization. In the abstraction phase significant relations are selected. Construction of the model is followed by analysis of the model, which leads to certain conclusions; this is known as deduction. After this, these conclusions must be translated into verifiable statements concerning the original system; this is known as realization.

Realization consists of two parts: validation and implementation. Validation verifies the conclusions in order to find out whether the model is valid. If there is no falsification, we can indeed implement the conclusions from the model (implementation). If the results of

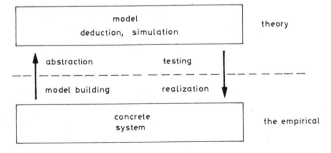

Figure 24. The model cycle.

validation do not satisfy the criteria of the test, we can start the model cycle again. In this process good use can be made of the information obtained in the first model cycle.

Analysis of a concrete system is often not a well-ordered process. Intuition and feeling play a major part. We do not regularly follow the precise sequence of abstraction, deduction and realization. We start with a rough model of the system and refine it in subsequent model cycles until it fulfils the criteria. This is known as model refining.

6.5. MODEL CLASSIFICATION

6.5.1. Introduction

We can classify models in various ways; from very rough schemes to very detailed ones. Rosenblueth and Wiener [77] were among the first to note that a representation to scale and also a mathematical law both belonged to models. Their first subdivision of models consisted of:

1. 'Material models', transformations of original physical objects, the representation of a complex system by a different physical system, which is assumed to be simpler than the original system and yet resembles it.
2. 'Formal models', symbolical statements in logical terms about an idealized, relatively simple situation, the statements presenting the structural characteristics of the original factual system.

The idea of model classification is that it creates classes to which we can attribute general characteristics. A researcher using a specific model for solving a problem can then use the general characteristics of the class his model belongs to. He will then also be more readily aware of the limitations the use of such a model involves.

Mihram [70] tried to classify models on the basis of the systems for which models have to be made. First, he differentiates between static and dynamic systems. Then he divides these two categories into deterministic and stochastic systems. He divides the models into 'material' and 'symbolic' models, which in turn are subdivided into 'replication', 'quasi-replica' and 'analogue'. If we put this classification in a matrix with the subdivision of the systems on the side, and one of the models on the upper side, then we have created a typology of models.

In our view, this classification is based on the kind of model that is used. However, we can also distinguish between the method or working principle of the model and its function. We believe that Bertels and Nauta [72] give a more complete survey of model classifications. It is based on the following:

1. For what kind of system is the model used: concrete, conceptual or formal?
2. What is the method or working principle of the model? The most important kinds of models are: scale models, analogue models, ideal models, structural models, mathematical models and abstract models.
3. A classification according to the model's function, distinguishing between six main functions:
   - models with explorative and heuristic functions;
   - models with descriptive and reductive functions;
   - models with explanatory functions;
   - models with operationalizing functions;
   - models with formalizing functions;
   - models bridging gaps between abstract theories and concrete applications.

In order to show the possibilities and applications of models, we will discuss Bertels and Nauta's classification in more detail.

6.5.2. *Typology of models based on the nature of systems*

Systems can be divided into three kinds: concrete, conceptual and formal. In this:

concrete refers to   *matter*,
conceptual to ⟶ *concept*,
formal to   ⟶ *abstract name*.

A model is a system which corresponds to a certain extent to another system. We can classify models as concrete or empirical; conceptual, divisible into theoretical and realization models (a division corresponding to the division of science into empirical and formal sciences); and formal.

Concrete (or empirical) models

Empirical models can be models of concrete, conceptual and formal systems. For example:

1. An empirical model of a concrete system, for instance, is a planetarium as a model of the solar system.
2. An empirical model of a conceptual system (we could call this an application), for instance, is the pyramid of Gizeh as an application of the stereometrical figure known as a 'pyramid'.
3. An empirical model of a formal system (we could call this an applied realization model), for instance, is the ruler as an applied realization model of the formalized theory of real numbers.

Conceptual models

Conceptual models are perhaps the most frequently used; they are systems with which we try to describe empirical reality. Examples of conceptual systems are: number systems, line and point systems, networks, figures, patterns, drawings or the periodic system of the elements. All these systems can, of course, be used as conceptual models.

1. A conceptual model of an empirical system, such as a technical drawing of an existing house, as drawn up after measurement, or a map of a city.
2. A conceptual model of a conceptual system: an architect has a certain image of a house which he wants to design. This image can be regarded as a conceptual system. From this conceptual system the architect will have the engineer make a conceptual model in the form of a technical drawing. The next step will be to build the house itself with the aid of this conceptual model (the technical drawing).
3. A conceptual model of a formal system: a formal system, for instance, is the abstract form of language in which a conceptual system is formally represented by symbols for the purpose of study. Further examples of formal systems: the language systems of mathematical logic and, the best known: the axiomatic systems of mathematics, in so far as no use is made of the meaning of the formulae. One must read, for example, A is greater than B (with an arrow to the right), as: between A and B there is a certain order relationship. The circuit diagram of an And-circuit in a pneumatic computer can be looked upon as a conceptual model of the logical formula $(A + B)$.

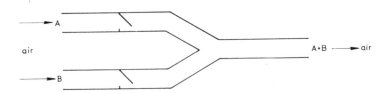

Figure 25. Circuit diagram of And-circuit in a pneumatic computer.

Formal models
In the foregoing we have already tried to make it clear what formal systems mean. Each formal system itself can be used as a formal model. Hence we can again distinguish between formal models of empirical, conceptual and formal systems.

1. A formal model of an empirical system. As a rule we will make conceptual models of empirical systems. From these models (which are also models of other conceptual systems) we can then make formal models. Actually, these formal models are again models of the empirical system, but they are also models of conceptual systems. For example, we can define a certain characteristic of an automobile (motion) with a linear differential equation. This can be regarded as a conceptual model of the automobile. From this differential equation we can again make a formal model by replacing the various coefficients by meaningless letters. This can then be regarded as a formal model of the concrete system: the motion of an automobile. A second example is an uninterpreted mathematical model of an economic process.
2. A formal model of a conceptual system. We have already given an example of this in 1 above.
3. A formal model of a formal system. For this we can go to analytical geometry, which shows clearly that algebra is a translation of geometry and vice versa. We can thus use algebra as a formal model for the formal system of 'geometry' and vice versa.

6.5.3. Typology of models according to function

This classification is confined to several examples.

1. Models with an explorative and heuristic function. They are often used for formulating a new theory or broadening an existing one. The function of the model is often the function of a working hypothesis. This working method has already been used frequently in seeking scientific knowledge. Electrical phenomena have long been explained by analogy with the flow of water.
2. Models used for description and reduction. These are used when the reality is so complicated that a complete description becomes pointless. An example from corporate management: the organization chart as a model for organizational structure. The reality is far more complicated than such a simple model can represent.

Yet such an organization chart is useful in describing an organization, explaining it to others, and so on.

3. Models with an explanatory function. In this case we can use the same model as for the descriptive and reductive functions, viz. the organization model. We can also explain certain phenomena and events in the organization with the aid of this organization chart. Analogous models can generally be used for this purpose. For instance, by regarding the human heart as an engine or a pump, certain phenomena can be explained, such as death following a heart attack.

4. Models with an operationalizing function. For instance, scale models which can be used for conducting experiments, such as an airplane model in a wind tunnel. The reduced model enables us to carry out experiments which would be difficult with the system itself. Other examples are computers which simulate certain characteristics of the nervous system, or models with electronic components, for instance of the flow of rivers.

5. Models for formalizing and automating research. An example is a mathematical model of the airplane model mentioned above, which can be experimented with by means of calculations. In this way, the actual experiments can be dispensed with. Simulation models for business enterprises can also be regarded as such.

6. Models used for mediating between theory and reality and simulating behaviour. An example is the model that was made to simulate the behaviour of the belligerents in Vietnam.

6.5.4. Typology of models according to method or working principle

This is perhaps the simplest classification, based on the working principle of the model. A summary may suffice:

1. scale models;   4. structural models;
2. analogous models;   5. mathematical models;
3. ideal models;   6. abstract models.

One of the most important ideal models, the 'black box', is discussed in 6.7. In 6.6. we pay special attention to mathematical (quantitative) models.

6.6. QUANTITATIVE DESCRIPTION OF SYSTEMS

6.6.1. Introduction

In order to arrive at a description of a system, it is essential to have a number of auxiliary concepts. Before we deal with these, we must first ask what the description of a system implies. In essence, it must be adequate for visualizing the system in one way or another. We make a model.

A description or a model of a system is a representation of the actual system. For the various magnitudes in the system we need substitutes in our description or model. In practice we often find magnitudes in our system to which we attribute numbers or quantities which we arrive at by measurement. These will then function as substitutes for the real magnitudes. In describing these magnitudes or changes in them we use variables and parameters as substitutes. In this way, we create models in the form of sets of equations, known as quantitative or mathematical models.

It will be clear that many quantitative descriptions (quantitative models) of systems are possible. We will not summarize these models, but will give only a brief description of a meta-model, a model of the quantitative models. Such a meta-model is the system cell [76].

6.6.2. System cell [76]

A system can be described as a complex of variables and parameters. The most important are the variables, since they impart a dynamic character to the entities of a system. We can divide the variables into five types:

1. Input-variables. These are the representations of magnitudes carried into the system from the environment and on which the system is considered to have no influence.
2. Decision or control variables. These are internal variables which can be manipulated and which enable us to choose from the available alternatives by reference to certain criteria. In fact, we can control the system with them. We decide, by means of these variables, what will happen to the input of the system.
3. Auxiliary variables. These are dependent variables introduced into system equations in order to make it easier to find a solution. They can be eliminated at any time, though this often makes the equations more complicated.
4. State variables. These are representations of state magnitudes. They refer to the system's momentary internal situation; they are the consequence of the preceding input and decisions, but cannot be directly manipulated. All relevant historical information is assumed to be represented in the state variables (see 2.6.). State variables, together with the other variables, are related in functions from which the output variables at time t can be deduced. The behaviour of state variables in terms of time depends on the state variables themselves and the values of the independent variables.
5. Output variables. These are determined by input and internal system variables. What constitutes an output variable is determined by the problem with reference to which the system is studied. Examples of output variables are earning capacity, generated energy and job satisfaction.

Next, there are parameters. These are the system magnitudes not included in the variables. They are sometimes referred to as mappings of characterizing magnitudes. In fact they define the relations between input, output and system variables. An example is the mains voltage and type of current (A.C. or D.C.) which we can call the parameters of an electromotor. These concepts make it possible to discuss the system cell. 'It is a scheme of a system (fig. 26), whereby

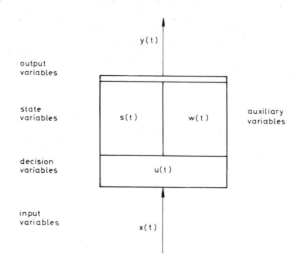

Figure 26. System cell.

the variables are listed according to the classification given for it' [75]. Thus we have a model for mathematical models, a meta-model.

This arrangement is sometimes used for classifying systems. Systems can then be characterized with a 'quartet' k, l, m, n, consisting of the number of input variables k, decision variables l, state variables m and output variables n. It makes it possible to describe different systems of the same class (the same quartet) by analogy, for example the electrical-mechanical analogue discussed in 6.3.

6.7. BLACK BOX

The 'black box' is an ideal model. Its importance is that it actually indicates a method of research. In this method either we have no interest in the structure of the system, or it is not known. We then try, by studying input and output, to learn about the behaviour of the system and the structure that may cause this. We see that the researcher together with the system being studied (the black box) forms a super-system. Together with the black box, he forms a system with feedback (see 7.4.2.). For the researcher the black box inputs and outputs are observable and clear. These factors link the researcher with the system. The openness of a black box is therefore limited, because the interaction between system and environment is predetermined.

The term 'black box' arose in electrical engineering (circuits used to be moulded in black bakelite). When a black box in a system consisting of several such boxes broke, the technician had to trace the defect with a black box approach and replace the box concerned.

Briefly, therefore:

> 'a black box is a system whose contents are unknown to us or do not interest us, and whose relation with the environment is predetermined'.

By viewing concrete systems (systems in reality) as black boxes, we can describe them functionally and clearly and study them experimentally, without the risk of damaging the system by opening it. Among other purposes, this is very important for brain examinations and the like.

In general, it should be added that black boxes with the same external behaviour may have a different internal structure. For example, a given effect of an electronic system can be obtained with several different circuits. In the black box approach this leads to the theorem of the undetermined structure or, in other words, a given behaviour may be obtained with an infinite number of structures. With the black box approach we can make assumptions regarding the possible internal structure, but cannot give a decisive answer without opening the box.

A researcher wanting to examine a system with a black box approach proceeds as follows. He selects various inputs, and records the

resulting outputs. He can manipulate the input, and by studying the
changes in the output he can learn about the behaviour of the system.

The researcher must, however, observe a number of conditions
[78], which can be generally summarized as follows:

1. During the experiment the black box must be completely insul-
   ated so that other stimuli (inputs) that have an impact on the observed responses (outputs) cannot enter the black box. In certain
   cases, for instance biological systems, it is difficult to comply
   with this condition.
2. The researcher must choose an environment in which the black
   box can be suitably manipulated.
3. During the experiment the observer must keep a 'protocol' of
   stimulus-response pairs in order of their occurrence. For systems
   with continuous behaviour this may be a graph of stimuli and
   responses. Guided by this protocol, he can try to find the rules
   governing black box behaviour. He therefore makes a description of this behaviour.

An illustration is a black box model of man [72]:

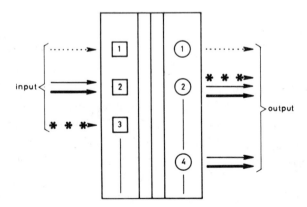

Figure 27. Primitive black box of man.

| input organ: | 1 = eye | 2 = mouth | 3 = ear |
| output organ: | 1 = eye | 2 = mouth | 4 = anus |

protocol:

input:	output:
light (1)	pupil reflex (1)
food (2)	saliva and faeces (2 + 4)
food and bacteria (2)	saliva and diarrhea (2 + 4)
words heard (3)	words spoken (2)

6.8. RESPONSE STUDIES

6.8.1. Response

Response is defined as the system's reaction to a specific input or change in input, measured by the output. This relates to the characteristics of the output of a system after receiving a specific input. This concept is widely used in studying systems by the black box approach or input-output analysis. The procedure is as follows: a fully specified input is introduced into the system and then the response of the system based on the variation in output over a period of time is studied. In practice, this is done by making a protocol of series of inputs and the subsequent outputs (6.7.). This makes it possible to construct an input-output model which describes the response of the system to a specific input (see 6.7.). This is used to try to draw conclusions about the structure and state of the system. Examples are given in 6.8.3.

This research approach has been successful in many sciences. Sometimes, for example in neuro-psychology concerning brain research, any other method is impossible. This applies to research into objects that cannot be 'opened' without substantially changing the character of the system or sometimes even making it uninteresting for further research in that science.

6.8.2. *Linearity* [79]

Models of systems are often subdivided into linear and non-linear models. For systems from reality this characterization is irrelevant, but it offers many advantages in describing systems. We often abstract from reality to such a degree that a system can be described with the aid of a linear model.

Linear systems are characterized by two qualities: homogeneity and additivity. Homogeneity means that if we change the input by a factor K, the output also changes by a factor K.
  As a formula:
if $x(t)$ is the input and $y(t) = f(x(t))$ the output,
then the system is homogeneous if the following applies:
input $K \cdot x(t)$ gives output $K \cdot y(t)$, hence:

$$K \cdot x(t) \rightarrow K \cdot y(t) = K \cdot f(x(t))$$

The system is additive if the output for two inputs is equal to the sum of the outputs the individual inputs. Hence the input is $x_1(t) + x_2(t)$; the response to it is $f(x_1(t) + x_2(t))$; then the system is additive if the following applies:

$$f(x_1(t) + x_2(t)) = f(x_1(t)) + f(x_2(t))$$

For systems with a memory this may cause difficulties. Here the state variable plays a role. We then call a system linear in relation to a certain initial state. A system is not linear if it does not satisfy one or both conditions of homogeneity and additivity.

Differentiation into these two classes is important because linear systems can be described mathematically with linear equations (mathematical model). For a given input one can arrive at an analytical solution of the system equations. The solution of these equations is then the response of the system to a received input signal. Stability research of systems is easier this way. Because of these convenient qualities of linear systems, one will often attempt to represent systems which occur in reality with models based on the assumption that systems behave linearly.

6.8.3. Examples of responses

This section deals with some of the more frequent types of responses. Though based on simple systems combined with simple inputs, the examples we have chosen are so general in their nature that even in more complex systems and less simple inputs we can often recognize these types of responses, especially if we assume the system to be linear. This can help in constructing models for describing the behaviour of such systems. Since the response of the system depends on the system and the type of input, we will confine ourselves to examples of systems which can be described by linear differential equations[1] and to a system characterized by a finite time delay (pipeline effect)[2], and to a choice from the four existing standard types of input: jump, pulse, sine and noise[3], viz. the pulse and the jump.

1. The order of a differential equation is defined by the 'highest' differential quotient appearing in the equation: d^2y/dx^2 as the highest differential quotient in an equation gives a second order equation; d^ny/dx^n gives an n-th order differential equation.
2. A system characterized by a finite time delay is formally described by an infinite order differential equation. This finite time delay is also called the 'pipeline effect', a name which suitably characterizes the way the system works. An ideal pipeline is a system which delivers the introduced input after some time without loss as an unchanged output. A system with a finite time delay does something similar: an input, offered to the system at a certain time, is delivered after some time by the system without loss as an unchanged output; this is where the term 'pipeline effect' comes from.
3. In response research into systems there are four 'standard' inputs: (ideal) pulse, jump, sine and noise.
a. An (ideal) pulse can be compared to a hard, very quick kick against a table, with which a great force is applied to the table for a very short time. In response research we observe the motion (response) of the table as a result of this kick. An ideal pulse is a kick which is infinitely hard applied (with infinitely great force) to the table for an infinitely short time.
b. An (ideal) jump can be compared to pushing against a table at a certain moment and with a constant force. Simultaneously, we study the movement (response) of the table.

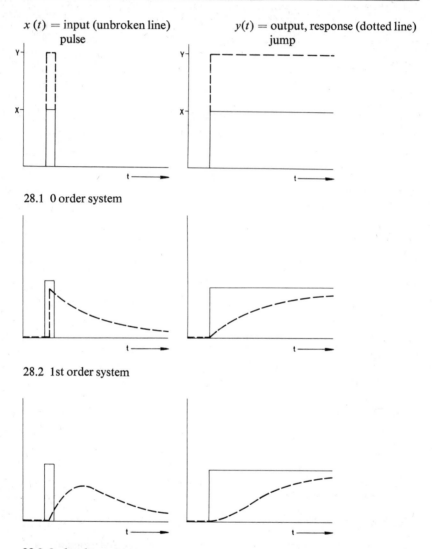

$x(t) =$ input (unbroken line)
        pulse

$y(t) =$ output, response (dotted line)
        jump

28.1 0 order system

28.2 1st order system

28.3 2nd order system

c. Sine can be compared to constant, frequent pushing and pulling of a table in such a way that each time a certain value of the force is reached, but not exceeded. The resulting movement of the table is again the response of the table.
d. Noise can be compared to arbitrarily pushing or pulling a table with a certain force which does not exceed a certain limit. We then study the table's response.

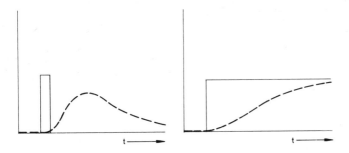

28.4 3rd order system

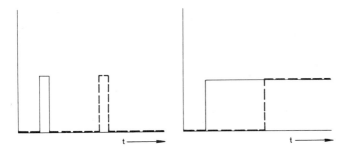

28.5 system with finite time delay ('pipeline')

Figure 28 continued. Examples of responses.

We shall discuss in greater detail the response of second order systems, described by:

$$a.\frac{d^2y}{dt^2} + b.\frac{dy}{dt} + c.y = x; y = y(t); x = x(t)$$

The behaviour of systems can often most conveniently be described with 2nd order differential equations without detracting too much from the exactness of this description.[4] These systems show a re-

4. 3rd and higher order differential equations correspond in their responses to those of 2nd order systems (see figs 28.3 and 28.4). The choice of description by a 2nd order differential equation is therefore often based on a compromise between simplicity and exactness of description.

sponse to a jump whose form depends on a certain relation Q, established by the coefficients a, b and c of the differential equation:

$$Q = \sqrt{(a \cdot c)}/b$$

Hence if $a = 8$, $b = 4$ and $c = 2$, then $Q = \sqrt{16/4} = 1$.

The response of a second order system is depicted below for different values of Q.

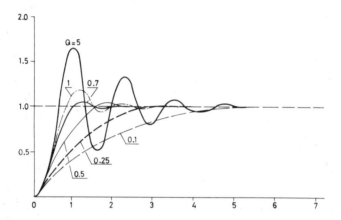

Figure 29. Response of a second order system for different values of Q.

Based on the response of a second order system we can estimate the magnitude of Q, also called the quality factor. It is possible to alter the value of the coefficients by modifying the system. This results in a change in Q presenting the possibility of designing systems with the response qualities we desire. A system's response is determined by the value of the coefficients which depends on the system parameters. By changing these we can construct a system with the desired response.

Apart from these specific forms of response of 2nd order systems to a jump as shown in fig. 29, other forms, all depicted in fig. 30, can occur as well.

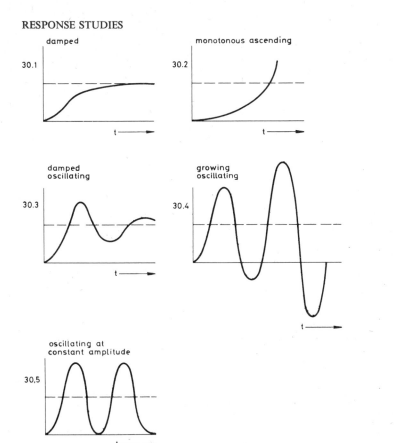

Figure 30. Existing forms of responses of 2nd order systems.

How do these examples of responses clarify the behaviour of systems existing in reality? Let us try to illustrate this with a simple example taken from Forrester [80].

It consists of a production and distribution system comprising a factory, factory inventory, wholesaler's inventory, and retailer's inventory, linked by order and product flows (fig. 31).

Starting from a steady state in all sections, Forrester illustrates with the aid of his 'systems dynamics' method the response of the various subsystems to a 10% jump in retailer's orders. As a result, the factory's production per unit of time behaves like the response of a

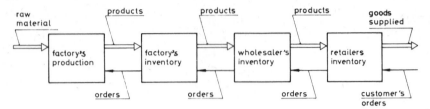

Figure 31. A production and distribution system.

system that can be described by a 2nd order differential equation. This implies that such an equation may be a useful model for an initial approach to the behaviour of such factory production.

As fig. 32 shows, the resulting variation in production level depends on the frequency with which the factory's inventory is adjusted. The lower this frequency, the greater the amplitude of the oscillation in the magnitude of production will be.

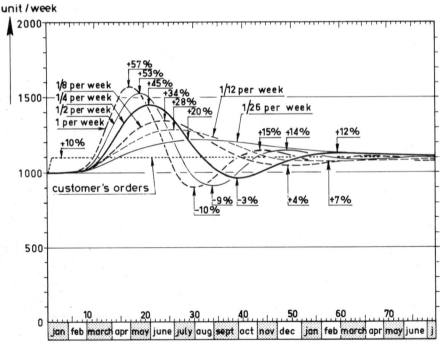

Figure 32. Changes in production level with a 10% increase in the number of orders received for different frequencies of adjustment of factory inventory.

6.9. BOULDING'S SYSTEM HIERARCHY

Instead of a typology of models, a classification of real-life systems can be used. Models are by definition also systems and therefore fall under this classification as well. Analysis of such a classification concerning the position of models in relation to empirical systems may give us an insight into the degree to which science is capable of describing reality with models. The most important attempt to do this was made by Boulding [81]. Van Peursen [82] has also worked on such a classification. We shall discuss Boulding's system typology in greater detail (see also Keuning [83]).

The economist Kenneth Boulding (one of the founders of the Society for General Systems Research) is one of the scientists who has tried to develop the general system theory. In 1956 he noted that an interdisciplinary approach, as advocated by the general system theory, can easily degenerate into a non-disciplined approach to problems. He regards the main task of general system theory as the development of a 'certain framework of coherence'. He then gives two possible approaches for structuring general system theory. The first is observation of the empirical world from which we can isolate several general phenomena found in various disciplines. These general phenomena are then used to attempt to construct general, theoretical models relevant to these phenomena. His second approach is to classify the areas or systems for study into a hierarchy according to the organizational complexity of the various individuals of which these areas or systems consist. Subsequently an appropriate theoretical description can be developed.

The hierarchy Boulding developed is the result of this second approach. In this hierarchy he arranges the various systems according to nine levels of complexity roughly corresponding to the complexity of the individuals or objects from the various research areas. He arrives at the following theoretical differentiation on the basis of his second approach:

Level 1 static structure, the framework level; a map is a good example. A new area of research is usually approached by using descriptions (models) based on this level.

Level 2 simple dynamics systems with certain necessarily pre-scribed movements. This can be called the clockwork level. Seen from man's point of view the solar system is a good example. Others are: the clock and simple machines. It is striking that time starts to play a role regarding the state of the system.

Level 3 cybernetic systems or control mechanisms. It is also called the thermostat level. Essential in these systems is the trans-mission and interpretation of information. As a rule they are systems in which we attempt to reach a desired value or norm by means of feedback. The essential variable in such dynamic systems is the difference between observed and desired values. If a difference appears, the system tends to react so as to decrease the difference. The best known example is the thermostat, which attempts to main-tain the temperature to which it is set.

Level 4 self-maintaining systems. It maintains itself by con-tinuous interaction with its environment. According to Boulding, this is the level at which life distinguishes itself from dead matter. It is also called the level of the cell. The cell is capable of infor-mation-transmission. One way in which it does so is self-dupli-cation in reproduction.

Level 5 is characterized by the plant. These systems have a division of labour. In one way or another the whole is divided into different functions. The plant has roots, leaves, etc. Certain cells have dif-ferent functions from other cells. But there is still the quality of 'equifinal growth' or prescribed growth; 'blueprinted growth'.

Level 6 is that of the animal. It is characterized by great mobility and awareness of its own existence. Another characteristic is that there are special organs for introducing information from the en-vironment. This system has an image of its environment.

Level 7 man who, besides the above, has the capacity of abstract thinking. He is able to use symbols. Besides an image of the world around him, he also has an image of himself. Man is probably the only system that knows it will die.

Level 8. At this level we find social systems such as organizations.

Level 9, the transcendental system. It is used by Boulding to put a roof on his hierarchy. Here he can place systems that do not fit into the other categories.

A characteristic of this classification into levels is that each successively higher level embodies all the lower levels. The consequence is the creation of a hierarchy embracing the whole of the universe.

What is the purpose of this classification into a hierarchy of systems? The important advantage, says Boulding, is that it gives us an impression of the gaps existing in scientific knowledge. Adequate models evolve from the first, second, third or at the most the fourth level. By detecting these gaps by means of such a classification it becomes possible to direct research towards these blank spots. In this way general systems theory can be of assistance in organizing these efforts. In order to study these blank spots we use models. As an initial step we generally use models of a lower level to study phenomena corresponding to higher levels.

At the first level we find adequate descriptive models for practically all the different disciplines. An example is the use of an organization chart in administrative sciences. When we do this, we use a first-level model to obtain an insight into an eighth-level system. This may be a very useful procedure. Boulding's classification makes us realize that using a first-level model in order to obtain an insight into eighth-level systems involves certain risks. We omit a number of relevant aspects from our observation, and we should fully realize the risks involved. The traditional model of an enterprise in economics, in which labour, resourves and energy are transformed into products, is an example of a second-level model which is also used for an eighth-level system. The principle of budgeting and budget-control is an example of a method based on a cybernetic model of the managing process of the enterprise. This is a method derived from the third level applied to a system of the eighth level.

6.10. REFERENCES

70. Mihram, G. A., 'The modelling process', *IEEE transactions on systems, man and cybernetics*, 2, no. 5, november 1972, pp. 621-629.
71. Bertels, C. P., 'Het model op de wip', *Intermediair*, 8, no. 11, pp. 23-25.
72. Bertels, C. P. and D. Nauta, *Inleiding tot het modelbegrip*, Bussum 1969.
73. Apostel, L., 'Towards the formal study of models in the non-formal sciences', *Synthese* 12, 1960, pp. 125-161.
74. Leeuw, A. C. J. de and W. Monhemius, *Methodologie en inleiding systeemleer* II, syllabus Eindhoven University of Technology, 1973.
75. Ashby, W. R., *An introduction to cybernetics*, New York 1956.
76. Hanken, A. F. G. and H. A. Reuver, *Inleiding tot de systeemleer*, Leiden 1973.
77. Rosenblueth, A. and N. Wiener, 'The role of models in science', *Phil. of Sci.*, 12, 1945.
78. Klir, J. and M. Valach, *Cybernetic modelling*, London 1965.
79. Zadeh, L. A. and C. A. Desoer, *Linear system theory*, New York 1963.
80. Forrester, J. W., *Industrial dynamics*, Boston 1961.
81. Boulding, K. E., 'General system theory – the skeleton of science', *General Systems* I, 1956.
82. Peursen, C. A. van, C. P. Bertels and D. Nauta, *Informatie*, Utrecht 1968.
83. Keuning, D., *Algemene systeemtheorie, systeembenadering en organisatiekunde*, Leiden 1973.

CHAPTER 7

CYBERNETIC SYSTEMS

7.1. INTRODUCTION

In this chapter we shall discuss a particular group of systems: cybernetic systems. In order to get an idea of the kind of system we mean, we can refer to Boulding's typology (6.9.). Boulding [84] puts cybernetic systems at the third level. He says: 'The next level is that of the control mechanisms or cybernetic systems, which might be nicknamed the level of the thermostat'. The main difference between third-level systems and those of the two preceding levels is that in cybernetic systems the transmission and interpretation of information play an essential role. Boulding says (1956) that there has been a great development in the area of study of this level, and that the theory of 'control mechanisms' has developed as a new discipline called cybernetics.

The word 'cybernetics' comes from the Greek word 'kybernetes' (steersman). The Greek philosopher Plato used it in his discussions about the analogy between navigating a ship and governing a country or group of people. In 1840 the term was rediscovered by Ampere in his classification of the sciences [85].

Early in history, principles from cybernetics were frequently applied without cybernetics being explicitly mentioned. We find the first applications in Arabic and Greek manuscripts around 200 B.C. [86], where control systems are mentioned.

Other known instances (in chronological order) are: Archimedes' automatic waterlevel-regulator for waterclocks; the on-off control as the 'escapement of the mechanical clock', designed in China in the Middle Ages; the regulation of the grain supply for a flour-mill by the Frenchman Ramelli in the 16th century, and Cornelius Drebbel's thermostat. (1593-1633) [87].

The most important technical and social application of a regulating mechanism is 'Watt's Governor', a mechanism developed by James Watt in 1788 regulating the number of revolutions of a steam-engine independently of the load. For the first time it became possible to apply the steam-engine on a wide scale.

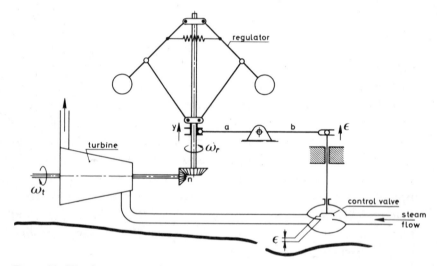

Figure 33. Watt's governor.

The principle of Watt's Governor was worked out mathematically by Maxwell in 1868 [86]. Its operation is based on negative feedback. This principle was applied in 1873 by Farcot [86] in a steering mechanism for ships. The great technical breakthrough was when Nyquist and Black [88] in 1931 designed the electronic amplifier based on negative feedback.

What is nowadays understood by cybernetics? Wiener [89], the man who in 1948 gave the great impulse toward the theoretical development of this new discipline, defines cybernetics as: 'the science of control and communication in the animal and the machine'.

He says furthermore: cybernetics is not a science which studies systems, but a science which studies the behaviour of systems. It does not ask: 'what is this'?, but: 'what does it do'? Note that Wiener (the mathematician) in 1948 was not aware of the fact that he was actually dealing with the concept of cybernetics in the same way as Plato did.

Ashby [90], being a biologist, describes cybernetics somewhat differently from Wiener. He calls cybernetics the science which deals with 'the study of systems that are open to energy, but closed to information and control; systems that are information-tight'[1]. The fact that Ashby had done a great deal of research into the functioning of the human brain may have contributed to his definition of cybernetics. But Ashby, too, applies the principles of cybernetics to machines as well as people, and to groups as well as well as organizations.

The development of communication and information theory, especially the formal approach by Shannon [91] (5.2.5.), made a major contribution to the further development of this new science. Cybernetics, in solving control problems, could make good use of the results from information theory, since the concept of information plays a big part within these control problems.

Cybernetics has since become important in certain areas of science in particular. First, control engineering, where the automation of all kind of processes, based entirely on cybernetic principles, speaks for itself. Second, negative feedback is used frequently in biology, and also in a number of medical sciences such as neuro-physiology. Third, in communication and information theory, machines, processing data often use cybernetic principles.

---

1. Information-tight, literally 'water-tight' to information, refers to the fact that information contained in the system (stored in it) cannot be disturbed by influences from outside (noise, see 6.8.3) in such a way that after decoding, the original meaning is lost (i.e. that the information-content has been disturbed [90, s9/19]).

It is intersting to note that cybernetics has only hesitatingly been accepted in economics. But in recent years there has been faster progress in applying cybernetics to macro-economic planning for countries with a centralised economy [92].

On the whole it can be said that there are a great many different definitions and descriptions of cybernetics. From these many definitions, however, two general characteristics can be deduced. First, we examine teleological systems; second, we attempt to find generally valid rules which relate the behaviour and structure of such systems to their objectives.

We regard cybernetics as part of general systems theory. It is concerned with an important aspect of the behaviour of systems: control.

This chapter first deals with block diagrams (7.2.), in order to have an adequate model for representing cybernetic systems. After this, we discuss the concept of 'control' (7.3.); what we mean by this; what general conditions have to be met to obtain effective control. Special attention is paid to the 'law of requisite variety' formulated by Ashby [90]. In section 7.4. we deal with the differentiation often made within control system, viz. cause-controlled and error-controlled systems. In 7.5. we illustrate the preceding with some examples of 'control systems' from various sciences, in order to clarify the generality of the cybernetic concepts and to simplify their application.

7.2. BLOCK DIAGRAMS [94], [93], [95]

In order to understand the behaviour and functioning of cybernetic systems, three models are regularly used. First, we have the mathematical model, generally consisting of a set of equations; second, the block diagram; and third, the flow chart (2.6.). We will concentrate on the block diagram, since it forms an easy, visualized, functional representation of cybernetic systems. It is a simple model, useful for many purposes.

A block diagram is a graphical representation of cause and effect relations between the input and output of a system. It is a convenient means of casting light on these relations. The simplest way to draw a block diagram is a single block with only one input and one output (fig. 34).

Figure 34. The single block.

The block contains either a description or the name of the element, the system or its function, or it contains a symbol for the mathematical operation through which the input has to pass in order to become the output. Fig. 35 gives an example in which the input is x, the output is y and the operation represents differentiation of x to time.

Figure 35. Differentiation visualized in a block diagram.

Generally, the block diagram is very easy to use for determining the transfer function (when this is possible) $Y = H \cdot X$, where X represents the operation as executed by the element (the block); in the example in fig. 35, H is then equal to d/dt. As a rule, a block diagram in its simplest form looks like fig. 36.

Figure 36. The simplest block diagram.

Apart from blocks, a block diagram contains two more symbols;

symbols for deducting or adding magnitudes, and branching points. The adding and deducting points are represented as follows:

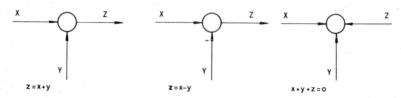

Figure 37a. Adding point.   Figure 37b. Deducting   Figure 37c. General repre-
                             point.   sentation of an
                                                      adding point.

The adding (or deducting) point is a general representation of an algebraic equation. The inputs coming to the adding point via the arrows are put in one part of the equation, with or without a minus sign. On the other side we put the outputs leaving the adding point via the arrows. In fig. 37a, x and y go towards the point: thus in the one part we have $x + y$; in the other we have only z because this goes away from the point.

A branching point is where one input supplies a larger number of identical outputs (fig. 38).

Figure 38a. Branching point.   Figure 38b. Branching point.

With the aid of these three elements, blocks, adding points and branching points, we can draw block diagrams and use them to construct adequate models for representing cybernetic systems.

In order to give an idea of how block diagrams can be translated into mathematical models, we now work out two illustrations.

First we take as a model a block diagram in which two or more blocks appear in sequence. In this case there are two possibilities: the first is that the process in the first block does not influence that in the other. The second is that the processes in the different blocks do influence each other. In the former case it is easy to work out the model (fig. 39):

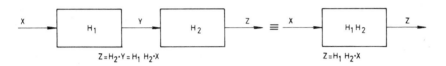

Figure 39. Sequencing of blocks.

But if the processes in the blocks do influence each other, the calculation mentioned above is no longer applicable. Multiplication of the two transmission functions H_1 and H_2 is not automatically possible. We quickly find ourselves faced with complex calculating methods (see, for instance, Nauta Lemke [94]).

A second example is when blocks appear in the form of a parallel connection. In this case we can find the common transfer function by adding the specific transfer functions together (see fig. 40):

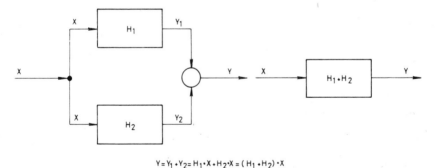

$$Y = Y_1 + Y_2 = H_1 \cdot X + H_2 \cdot X = (H_1 + H_2) \cdot X$$

Figure 40. Connecting blocks in parallel.

With the aid of these calculating rules and sign agreements we are now able in principle to represent any cybernetic system and other systems as well. In fact, we can now change or simplify block diagrams with these rules, as the above examples show.

7.3. CONTROL

Control is a behavioural characteristic of a system; it is not inherent. We can describe it as the directed influence upon another system, subsystem or entity through interaction.

Sommerhoff [96] distinguishes five characteristics of control:

1. A set Z of outputs that can be realized by a system: all possible responses of a system, desired or undesired.
2. A set G, a subset of Z, consisting of all the desired outputs of the system attainable by effective control.
3. The set R of behavioural possibilities of the regulator R^1 (for example management procedures).
4. The set S of behavioural possibilities of the system to be controlled S^1, resulting in an output of the system belonging to the set Z.
5. The collection D of disturbances influencing the behaviour (S) of the system and hence the behaviour (R) of the control system.

Hence the system has possibilities of behaviour resulting in desired outputs (belonging to G) and undesired outputs (Z/G). The behaviour S of the system is influenced by the behaviour R of the control system, so that the output of S will be part of G. This may be disturbed by events belonging to D, if R does not intervene adequately. Because of this, the output might no longer be part of G. A system that functions well under certain circumstances may not function well under others, it may be affected by the weather, varying demand, mechanical failures, all of which form part of D.

If the 'control system' or 'regulator' functions well (i.e. if, with its behavioural possibilities R, it influences the behaviour S of the systems in such a way as to result in an output belonging to the set G), there must by definition be a relation between D, S and R.

1. A system can only be observed by means of its behaviour. It is not necessary therefore to distinguish between the system and its behaviour. The symbol R will be used further for the set of possible behaviours and also to denote the control system. The same applies to the symbol S as regards the system to be controlled.

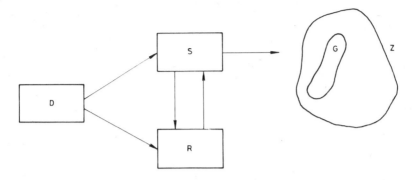

Figure 41. Control.

The following may clarify this. A typical example [97] of control is that of a hunter who shoots at a pheasant in flight. *D* consists of all the factors which introduce disturbances by the bird approaching from a particular direction. The disturbance may come from the hunter whose posture may be unsuitable at the crucial moment; from the wind which may blow from a different direction or from the light.

S consists of the hunter, with the exception of his brain, plus his gun. *R* is his brain. *G* is the behaviour resulting in a shot which hits the bird (the desired outcome).

R is now an effective regulator (it attains its goal) if, and only if, *R* relates to *S* for all values of *D* in such a way that their interaction results in certain behaviour which brings about an output belonging to *G*.

To guarantee this, the following conditions must be fulfilled:
1. *R* must have information about *D*.
2. *R* must have information about *S* (a relation with *S*).
3. *R* must have enough control measures.
4. There must be a criterion on which to base the choice of a control procedure by comparing the system's predicted behaviour with the goal.

Condition 1 is trivial. Without a knowledge of disturbances which may influence the system, adequate control will never be possible. It is the very influence of these disturbances on the desired output which is the object of control.

Condition 2 has more consequences, also for R itself. In working out this condition (formally) [97], it is found that R must contain a model of S in order to be an effective regulator. Without resorting to the formal derivation we can perhaps make this clear as follows:

R receives information on disturbances affecting S. In order to be able to cope with the effects of the disturbances, R must in fact predict the result of these disturbances upon the output which we want to be a part of set G. Since R cannot use S to predict these influences, R must have an isomorphic or homomorphic model of S with which it can predict the consequences of various control measures in combination with the observed disturbance, and can then choose the optimal measure based on the criterion (see condition 4).

Condition 3 refers to the 'law of requisite variety' [90], worked out by Ashby. Let us illustrate what this implies: Suppose D consists of five disturbances, 1, 2, 3, 4, 5. R has four actions available from which it can choose in order to obtain a desired given output in G. A given action by R produces a different output of S with each of the disturbances 1, 2, 3, 4 and 5. We also know what the outcome

		R			
		a	b	c	d
	1	A	E	D	C
	2	B	A	E	D
D	3	C	B	A	E
	4	D	C	B	A
	5	E	D	C	B

of the system will be given a certain disturbance and action. We can then, for example, draw up the following table (the outcomes appear as elements in the matrix A, B, C, D, E):

This then enables us to decide whether a certain desired outcome can be achieved whatever disturbance occurs (1, 2, 3, 4 or 5). Suppose B is the desired outcome. There are then no adequate measures against disturbance 1. If we want to make adequate control possible, supplementary measures will have to be taken, for example by adding a column e (new managerial actions). It is then possible to guarantee B as the outcome whatever the disturbances.

				R		
		a	b	c	d	e
	1	A	E	D	C	B
	2	B	A	E	D	C
D	3	C	B	A	E	D
	4	D	C	B	A	E
	5	E	D	C	B	A

Ashby [90] has deduced that the minimum number of different realizable outcomes (in integers) is:

$$\frac{\text{the possible number of different disturbances}}{\text{the possible number of different control measures}}$$

provided the condition is fulfilled that a given outcome can only be obtained by one control measure. For the above matrix this means, for example, that outcome A may appear once per column. Formulated in general terms: if we distinguish n different disturbances, we have to have a minimum of n different measures to guarantee a given outcome. In other words: Only variety in R can decrease variety in the outputs of the system as a result of D: 'only variety can destroy variety' (90, p. 207).

Condition 4 (the existence of a criterion on which to base the choice of the control measure by comparing the system's predicted behaviour with its desired behaviour) requires the existence of a goal or a number of goals (G), which must be part of the model which R has of S in order to be able to control S adequately. This means, in fact, that subset G must be defined with sufficient clarity within set Z. After they are established, these goals are incorporated in the model.

Some additional remarks seem to be called for. The primary purpose of control is to influence another system. This is done via interaction between the systems: regulator R and the system to be controlled S, consisting of information transmission. There is thus a direct connection between control and the subject of 5.2. Control does not refer to transmission of energy and matter, which may be essential for the system concerned, but to the information transmission which goes with it or is derived from it. We have argued that effective control requires a model of the system being controlled to be part of the control system. Since we aim at models that are only isomorphic with the system under observation, but can often use a homomorphic model, the scope of a regulator is generally limited.

7.4. CONTROL SYSTEMS

7.4.1. Introduction

Ideally, control systems can be divided into two main categories:

1. 'cause-controlled' systems [97].
2. 'error-controlled' systems [97].

Both are systems wanting to nullify the effects of disturbances on their correct behaviour. We call them both control systems. This division corresponds to two main methods of eliminating the effects of disturbances [99]:

1. The counter-measures are initiated by the disturbance itself:

this means that the state of the system is defined directly by the environment: cause-controlled systems.
2. The reaction of the system is defined by the difference between the actual and the desired value of the magnitudes which are to be controlled: error-controlled systems.

7.4.2. Cause-control

Cause-control is a form of control in which the control measure is defined directly by the disturbance itself. In order to make this possible, the regulator R must obtain direct information about the disturbances D, i.e. without an intermediary. (In 7.4.3. we shall see that this is different in error-control.) The working principle of a cause-controlled system is based on a cause and effect relationship.

Cause-control (and in 7.4.3. error-control) will be illustrated with a simple example: the control of the temperature in a room. It should of course be realized that we are dealing with a general principle applicable to a wide range of systems including social systems (Boulding's level 8: 6.9.). We can represent the system with the aid of the block diagram [99] of figure 42.

This cause-control system is designed to keep the temperature in the room at a chosen value which we wish to maintain (Z therefore

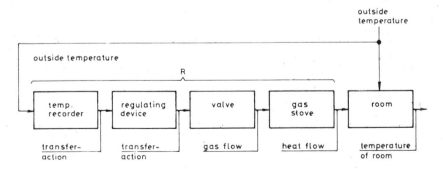

Figure 42. Cause-control of the temperature in a room.

consists of a set of possible temperatures; G of one or a set of desired temperatures). The temperature is brought to the desired level by the following series of activities: the outside temperature is measured with a recorder. When heating a room we shall usually assume that the outside temperature is lower than the desired room temperature. This influences the room temperature, because the heat from the room with its higher temperature leaks into the environment with its lower temperature. The outside temperature is a disturbance and belongs to set D. This disturbance influences the room and its temperature (the system S to be controlled) as well as the control system R consisting (schematically) of: temperature recorder, regulating device, valve and gas stove (fig. 42). The control system functions as follows: the temperature level is translated into a specific position of the gas stove valve. This determines the gas supply to the stove and consequently the heat flow to the room. Hence the room temperature results from the balance between the heat flow from the gas stove to the room and the heat flow to the environment through the walls.

If the outside temperature drops, the heat flow to the outside through the walls increases. This would mean that the room temperature will drop until a new belance is established depending on the heat flow produced by the stove. Before the drop is noticeable, however, the gas stove valve assumes a new position as a result of the drop in the outside temperature. This causes the heat flow to the room to increase. If everything has been correctly set, a new balance is created between incoming and outgoing heat flows which keeps the room temperature constant.

As already stated, in cause-controlled systems information on the disturbances is used directly for taking control action. On the whole, a cause-controlled system is based on the existence of a characteristic disturbance magnitude. This is measured and the resulting signal is used as a control magnitude for the process, in a number of successive steps or not. In control engineering one speaks of an open-loop system, in which there is no circular information transmission (i.e.

information leaving a certain point and returning to it later).

A general schematic presentation (block diagram) of a cause-controlled system is given in fig. 43.

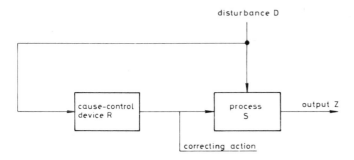

Figure 43. General cause-controlled system.

The characteristic properties of cause-controlled systems can be easily formulated with the aid of this figure:

1. The impact of the disturbance is counteracted before it is able to affect the magnitudes we want to control.
2. In order to be able to have effective control, all possible disturbances influencing the process must be known.
3. For a cause-controlled system to function properly it is essential for us to know exactly how a certain disturbance affects the system, so that we can adjust the cause-control correctly. This presupposes a model of the system. For instance, if we do not know the influence of a drop in the outside temperature upon the inside temperature, it will be impossible to decide the correct setting of the gas-stove valve.

Having regard to these characteristics, it will be clear that cause-control is often possible in relatively simple and fully determined systems.

7.4.3. *Error-control*

In contrast to cause-control, where we use a chain-shaped information-transmission structure, error-control employs a circular structure. This has direct implications for our discussion [100].

In cause-control we speak of a beginning (the input) and an end
(the output), and subsequent phases in transmission are determined
directly by the preceding phases as a result of a cause- and effect
relationship; in error-control we have neither a beginning nor an
end owing to the circular transmission .Statements such as: 'if *a*
then *b*' have little meaning if *b*, directly or indirectly, again yields *a*.
Circles have no beginning.

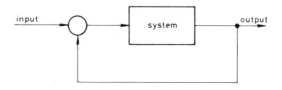

Figure 44. System with a circular information-transmission structure.

Such circular information transmission may be 'positive' or 'negative'. In both cases part of the system's output is fed back into the
system as information about the output. We speak of 'positive' (or
positive feedback) when the information carried back serves as a
measure of the deviation in output compared with a certain level,
and this information moreover gives a impulse towards increasing
the deviation. 'Negative' (or negative feedback) indicates that the
information about a deviation in output compared with a certain
desired level or norm is used to reduce the deviation.

This latter form, which we call error-control, will now be discussed. The foregoing also allows us to explain the term 'error-
controlled'. The deviation in the output compared with the desired
value acts as an initiator of the corrective action. The control system
R, however, receives information on disturbances *D* which affect
the system *S* to be controlled, only after the information has passed

through *S*. A general error-controlled system in terms of 'control' is as follows:

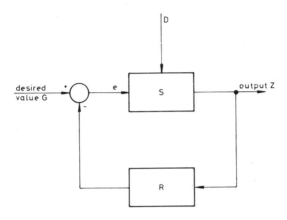

Figure 45. A general error-controlled system.

e is the error created by comparing the actual output with the desired value or norm. Thus set *G* (the desired output) consists of the desired value or norm produced as output by the system when $e = 0$. Set *Z* consists of all the outputs of the system. *R* will now try to influence *S* in such a way that *e* approaches zero, because this will produce an output in *G*. *R* will then be an effective error-control system if the error *e* approaches zero.

We can illustrate this with the example from the preceding section, controlling the temperature of the room with error-control instead of cause-control[1].

We start from a given desired temperature in the room, for instance 24 C. The actual temperature (say 20 °C) is now recorded and compared with the desired temperature (24 °C). There is thus a deviation (an error) of 4 °C. This signal (the difference) can be

1. Here again we are dealing with a general principle, applicable to a wide range of systems. The example is chosen mainly because it is comparatively simple and straightforward.

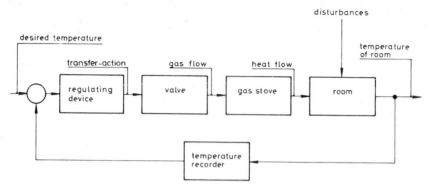

Figure 46. Error-controlled regulation of the temperature in a room (thermostat control).

transformed by an error-control device into specific action, which changes the flow of gas to the stove by means of a valve. As a result, the gas stove changes the flow of heat to the room, thereby decreasing the difference between the desired and actual temperatures. If the desired temperature is equal to the actual temperature, nothing more happens. This clearly shows negative feedback because information on the output is used as a means of decreasing the deviation from the desired temperature.

It is also clear what happens in the case of a disturbance. Suppose the outside temperature drops sharply. This will cause the temperature in the room to drop as well owing to increased heat loss through the walls. This creates a difference between desired and actual temperatures, and the error-controlled system will act as described above in order to maintain the desired temperature. Control will also clearly have been effective if the difference between desired and actual temperatures keeps on approaching Nil.

The most fully elaborated application of this general principle is found in control theory. In order to prevent confusion and misunderstanding, the concepts and nomenclature have been standardized by various bureaus of standards. In the Netherlands, the following general diagram for an error-controlled system has been arrived at;

in our view it could be generally applied provided an appropriate nomenclature for the more technical terms is established [93].

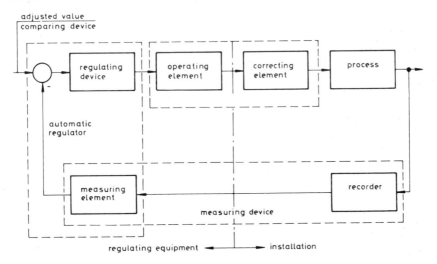

Figure 47. Block diagram of a general error-controlled system [93].

Error-controlled systems have the following characteristics:

1. They start functioning only when there is a difference between desired and actual values.
2. It makes no difference to their functioning whether the disturbance is known or unknown, expected or unexpected (as long as it remains within the range of error-control). In order to determine the appropriate control action, the system itself is used as a model.
3. They may be unstable (7.4.4.).
4. Independently of their state, they will still attempt to bring the actual value into agreement with the desired value. If, for example, someone opens a window in the room (intensifying the influence of the outside temperature), this will have disastrous results in cause-controlled regulation (with correction beforehand, not afterwards). This is not the case with error-controlled regulation.

As regards error-controlled systems it is also important to recognize the range within which error-control can be used effectively to influence the behaviour of a system. It indicates the system's limit in error-control. The error-control system cannot cope with signals falling outside these limits. This often gives rise to undesired effects such as instability. An example is the thermostat with a range of −25 °C to + 25 °C. Below or above these temperatures the stove either works at maximum capacity or turns off. (Many stoves in fact have an on/off regulator and not a continuous thermostat regulator.)

7.4.4. Stability of control systems

In practice, control systems, and especially error-controlled systems, show symptoms of instability. These are inherent in the circular information-transmission structure characteristic of error-controlled systems. There are two causes determining the response of these systems:

1. time delays in the system;
2. amplification of information in the system.

Before we deal with a practical example, we will first illustrate the influence of these two causes with a simple model [88].

The system consists of an amplifier with a variable amplification factor K (like an audio-amplifier) and a time delay of T seconds, i.e. an incoming signal is passed on T seconds after receipt. x is the in-

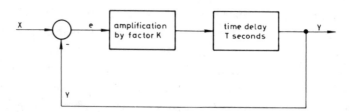

Figure 48. Demonstration model for instability symptoms.

put, y the output which is fed back; $e = x - y$ (the difference between input and output, the error). (This model has no practical significance. In practice systems are often more complex.) Fig. 49 shows the responses of the system for different K's and equal time delay T. The input is chosen in each case as drawn in fig. 49a.

This figure shows for $K < 1$, $K = 1$ and $K > 1$ respectively, the difference between signal e and the feedback signal y. The diagram is explained as follows:

Input x gives an equal difference signal e, since y is then still zero. e is amplified K times (this takes up no time) and then this amplified signal: $K \cdot e$, is held up for T seconds, after which it appears right away at the deduction point as y; x, however, has just become zero (x retains its value only for T seconds). So e then becomes $-y = -K \cdot e$ etc. The information keeps buzzing around in the loop. Per sequence it is amplified K times, i.e. after n times the circuit (nT seconds) y will be equal to $K^{(n-1)} \cdot x$.

Now it is immediately clear that the value of K influences the response. If $K < 1$ (fig. 49b and c), then y will decrease by a factor K with each successive sequence of T seconds. The system is then stable, as it returns to its original state after removal of the disturbance (4.5.). At $K = 1$ (fig. 49d and e) y remains equal to e, which remains equal to x. Nothing more happens. The signal keeps buzzing around at a constant amplitude. When $K > 1$ (fig. 49f and g) things get out of hand. The amplitude of y and hence of e keeps on increasing. The system behaves unstably.

It is clear that the amplification in the loop influences the stability of the system. As regards time delay, the best illustration is fig. 49d and e. The response of this error-controlled system at $K = 1$, i.e. no amplification, may not be unstable in the sense of fig. 49f and g, but the system keeps oscillating and does not return for good to a resting position. This is due to the time delay of T seconds in the circuit. Without this, there would have been no oscillation.

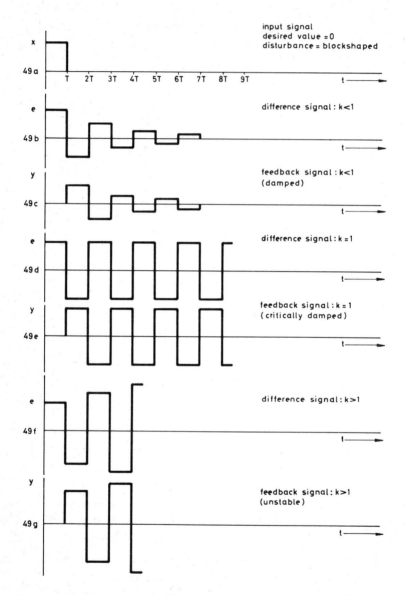

Figure 49. Response of the demonstration model at various K's.

The above example uses non-continuous signals. Systems which do have continuous signals as the input, such as sine waves or other continuously changing signals (this could be the above system), behave similarly. But the response will be continuous too. In reality, the output of a system cannot show infinite growth, as fig. 49g might suggest. Owing to limitations in the system, often of a physical nature, the output will at a certain moment have a constant (or practically constant) amplitude. Such a situation appears in radio broadcasts which include live interviews with listeners over the phone. It manifests itself as a sharp whistling sound, caused because the listener being interviewed has his radio on. This reproduces the conversation between him and the interviewer a fraction of a second later and makes it audible in the room. This sound is picked up by the microphone in the telephone. It is identical to the original signal, but delayed a fraction of a second. Through the feedback: telephone, transmitter, radio, telephone, with a time delay, the system becomes unstable and produces a whistling sound which keeps swelling until it can grow no more because of the physical boundaries in the system.

7.4.5. Combinations of cause-control and error-control

Closer examination of the characteristic features of both cause-controlled and error-controlled systems reveals certain differences. A cause-controlled system responds quicker. Control action is initiated before a disturbance can affect the system. Because of its structure it cannot become unstable. However, we have to know all the disturbances and what the effect of a certain disturbance will be in order to obtain effective cause-control. In error-controlled systems, however, we do not have to know in advance what the disturbances will be or how they will affect the system. Error-controlled systems may become unstable.

Since both systems have their specific advantages and disadvantages, and they complement each other on these points, it is logical to try and combine them (German: Störwertaufschaltung). Therefore a characteristic disturbance has to be singled out. In the example of heating a room by cause or error-control this is the outside temperature. The heat-loss from the room is determined mainly by the difference in temperature between the room and the outside. A combination of cause and error-control would provide good scope for effective temperature control.

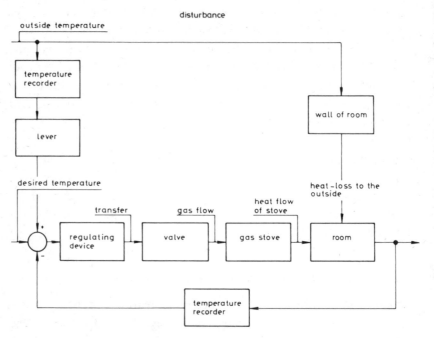

Figure 50. Control of the temperature in a room by combining cause and error-control.

Independently of error-control, the effect of the heat-loss from the room is anticipated by allowing for the influence of the outside temperature on this with the aid of the temperature recorder and lever. This signal is added to the difference (the error) between the desired and actual temperatures, and is used as information for intervening in the process.

We have tried to illustrate the combination of cause and error-control with this simple technical example. But it is applied in much more complex systems as well. An example [97] is the temperature of a cow's blood. This is regulated by the brain (error-control). If the temperature drops, the brain makes the cow's muscles and liver produce more heat, but the temperature has to drop first. If, however, icy cold air is blown at the cow, its blood temperature will go up without falling first. This is due to cold-sensitive spots on the skin stimulating the brain and causing it to take action (cause-control).

Perhaps this also illustrates that error-control always involves a deviation from the desired result. The goal will never be exactly reached. This is essential for an error-controlled system to function. A properly designed cause-controlled system will be able to reach the exact goal, provided the influence of all possible disturbances has been incorporated. This is because cause-control anticipates the effects of disturbances. If we can predict these effects exactly, we can also adapt the best counter-measures. In error-control this always happens after the event and the disturbance always influences the system first. Action is taken only if, as a result, the output starts differing from the desired value.

7.5. EXAMPLES OF CONTROL SYSTEMS

7.5.1. Introduction

This chapter has already given a number of examples of control systems. So far we have discussed simple models only, such as control of the temperature in a room. Many such instances can be found. There are also plenty of non-technical examples either of control systems or models based on cybernetic principles, but these are often more complex.

In the following sections we will deal successively with examples of technical control systems, man-machine systems, and models based on control systems in various sciences.

7.5.2. Technical systems

An example of a technical error-controlled system is Watt's governor discussed in 7.1. (see fig. 33).

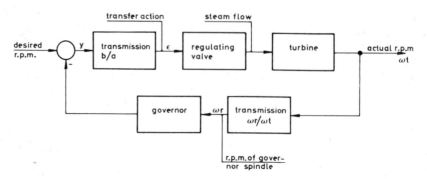

Figure 51. Block diagram of Watt's governor.

The desired r.p.m. is determined by the weight of the rotating balls.
The comparison between actual and desired r.p.m. is brought about
by the equilibrium between the weights of the balls and the centrifugal force with which the balls are flung out by the rotation of the
engine. The higher the r.p.m., the greater the centrifugal force. Consequently, the links pivoted to this rotating system move (transfer y),
operating the throttle, so that less steam is fed at a higher r.p.m. and
more at a lower r.p.m. In this way the engine is regulated at the
desired r.p.m.

   Another example is the automatic pilot in an airplane. This is
a relatively complex system designed to keep the plane to a certain
course in spite of atmospheric disturbances. It continuously com-
pares the actual course with the desired course, and adjusts the
position of the flaps, etc. according to the particular disturbance.

   Further examples are the systems known as servomechanisms,
such as steering machines in ships, power brakes in cars, automated
lathes, etc.

An example of a relatively simple cause-controlled system is an electric toaster [95]. The brownness of the toast is determined by setting a time-clock which fixes the length of toasting.

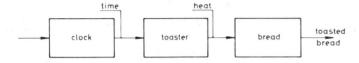

Figure 52. Block diagram of a toaster.

7.5.3. *Man-machine systems*

The controlling element in a control system is often an individual. One example is the toaster mentioned above, which is operated by an individual. If the toast is not brown enough, the time-clock is re-set and the toast will be darker. Other examples are the individual as the driver of a car, the pilot of an airplane, etc.

Such systems are illustrated by the example of the individual as a car driver [99] (fig. 53). He will keep on trying to adjust the car's actual direction, influenced by a variety of factors, to his desired direction. The same applies to the car's actual speed, which he adjusts to the desired speed.

7.5.4. *Control systems in various sciences*

In economics, the use of quantitative models has become very common. Some of them are based on cybernetic principles. This is not surprising, since economics also deal with the control of economic systems. A well-known problem is that of a national economy, for which many models have already been developed. In planned economies, as in countries with centralized administrations, models based on cybernetic principles are frequently used. An example is that of Romania, where the government is attempting to control the economy via a plan [92].

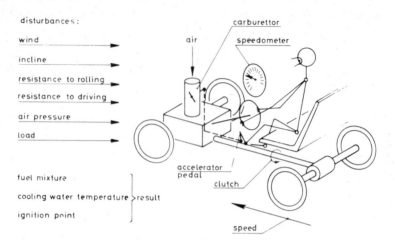

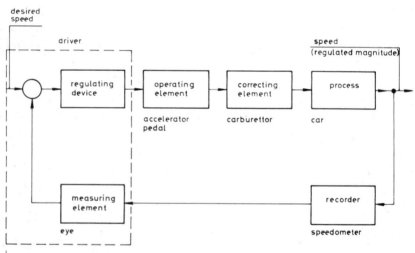

Figure 53. Speed control of an automobile.

Another example is that of Porter [85]. He says: Suppose the government of a country decides that its economy can be regarded as sound if a certain desired level of spending (S_d) can be maintained from month to month. If the actual level S_a is higher than S_d, there is an inflationary trend or an inflationary situation. In order to counteract this, the government asks the banks to raise interest rates. But if the actual level of spending S_a is lower than the desired level S_d, the economy will have to be stimulated by lowering interest rates. This apparently works out well if interest rates are raised or lowered pro rata to the difference between the actual and desired level of spending. Control of the level of spending can be illustrated with a block diagram:

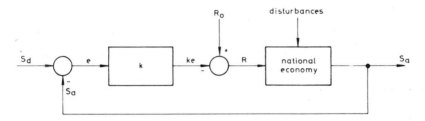

Figure 54. Control of the level of spending.

where

S_d = the desired level of spending
S_a = the actual level
e = the difference $S_d - S_a$

Furthermore, R (the rate of interest) = $R_0 + r$, in which R_0 is a standard rate of interest and $r = -k \cdot e$ is the change in interest depending on e, the difference between desired and actual levels of spending.

In the behavioural sciences such as social psychology, an example of a control system is found in the family, considered as a system (Watzlawick, [100]). Two types of feedback are observed. Positive feedback mechanisms are held responsible for change, growth and renewal, since they do not maintain a given state. Negative feedback mechanisms (known as family homeostasis) keep the family in equilibrium, since they contribute towards maintaining a stationary state. Watzlawick [100] gives examples of disturbed families greatly characterized by such family-homeostasis as a means of resisting the tensions created by the environment and other members of the family. This homeostasis may be so strong that families are practically immune to change and display a remarkable capacity for maintaining the status quo. Cases are described of families with schizophrenic children whose parents do not accept the signs of their children maturing and fight these 'abnormalities' by describing them as sick or bad.

Many examples of control systems can also be found in medical science, for instance in the human body. Two examples giving a good illustration of this are found in Wiener [89]. They both relate to human individuals who do not function properly. The first example is that of a patient entering a neurological clinic. He is not paralyzed and can move his legs if he gives the appropriate command. Yet he is handicapped. He walks insecurely with his eyes fixed on the ground and on his feet. He starts each step with a tug and throws his legs forward in turn. The second example is that of a patient who seems to be fine as long as he sits quietly in his chair. If he is offered a cigarette, however, his hand will reach out behind it. He will then make just as futile a movement to the other side and miss again. This will go on until finally he starts trembling violently.

In both cases the signal transmission which provides the feedback with which the movements are controlled, has been disturbed. The first patient's spinal cord is affected, stopping the signals from his feet, muscles, etc., which normally inform him of the state and position of these members, from reaching his brain. As his behaviour shows, he does not get the appropriate feedback through using his eyes.

The second patient has a disturbance in a part of his brain. The reaction by the muscles, representing the extent of the corrective action consequent upon a difference observed between actual and desired values, is not correctly adjusted.

7.5.5. Examples from business administration

Many examples are known that are based on cybernetic principles, since we find control problems of many kinds in business. Let us take four examples: inventory control [101], quality control [101], production control in the foreman/worker relationship [102] and cost control [101].

A business must have an adequate stock of good quality materials and finished products in order not to jeopardize the smooth progress of operations and sales. Every time a machine has to be stopped because there is no material in stock, or a sale has to be postponed or cancelled because of an inventory shortage, a business loses money.

On the other hand, it is important to stock minimal quantities since inventories can become obsolete and entail costs. On the whole, four magnitudes are important in inventory control:

1. the minimal quantity of a specific product that should be kept in stock;
2. the maximal quantity;
3. the re-ordering level;
4. the economic order-size.

Inventory control can now be regarded as a simple negative feedback cycle, with a desired inventory, an actual inventory and an error or deviation by the actual inventory from the desired inventory. On the bases of this information it is decided whether stocks should be replenished.

A limited definition of quality control describes it as ensuring that the properties of the product correspond to the prescribed standards and that these standards are well balanced. We do not mean, for instance, the work of a quality inspection department which measures, judges, and approves or rejects a product. We mean, rather, that the inspection results in corrective action to the production process such as fine readjustment of a machine.

The model which Ulrich [102] uses for foreman/worker relationships can easily be illustrated with the following figures. His method
of representation differs from that which we have used. But it does
reveal the essential characteristics of a block diagram with negative
feedback. His first example (fig. 55) is that of a worker/foreman/
manager relationship, in which the manager tells the foreman the
objectives or desired results (Sollwerte). The foreman tells the worker what to do and checks what is done in reality (sensor, measuring
device, or result measurer). On this basis, he can either correct the
worker or report to the manager. It is possible that the manager
changes or adapts his desired value as a result. Hence this system
can adapt itself.

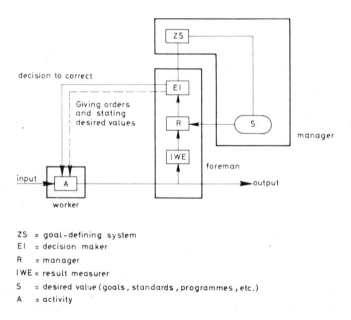

ZS = goal-defining system
EI = decision maker
R   = manager
IWE = result measurer
S   = desired value (goals, standards, programmes, etc.)
A   = activity

Figure 55. Worker/foreman/manager relationship [102].

In the second example (fig. 56) control is exercised by the worker himself; all he is given is the values of the desired results.

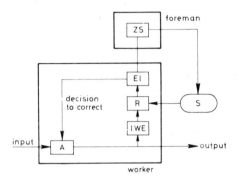

Figure 56. Transmission of control to the worker [102].

The third example (fig. 57) shows that in this way we can build up a hierarchy of error-controlled systems mutually influencing one another.

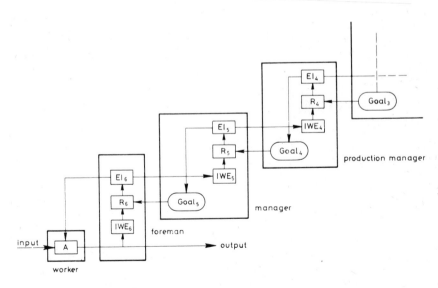

Figure 57. Hierarchy of control systems in a business [102].

Finally, we should like to discuss cost control as an example of a control system in business. Johnson, Kast and Rozenzweig [101] state that the aims of cost control are: to measure working efficiency and keep costs within certain limits. They say cost control involves: making a cost plan (standard costs), a means of measurement and comparison (cost accounting) and the corrective action of management in order to bring costs into agreement with the plan. Standard costs are carefully worked out estimates of basic costs. These estimates are based mainly on past performance data. Measurement of performance depends on working conditions, size and complexity of the business and assounting practices.

It is rarely possible to make a correct cost allocation for most operations. It is also impracticable to measure costs so exactly that the advantage of doing so would be wiped out by rising accountancy costs. When the data have been gathered they can be subdivided into units for comparison. Such a comparison would have to indicate every deviation from the plan. The management can then take various forms of action, such as: re-pricing the product, stopping production of a particular product, revising the advertising programme, starting a compaign to give the workers more motivation. If it were possible to allocate costs better, this might lead to improved control. Most cost-control problems arise from two factors:

1. problems of measurement in collecting data;
2. measurement data which are overtaken by events. These always relate to an earlier situation which may differ greatly from the actual situation. Decisions based on them are therefore liable to produce the wrong results.

7.6. REFERENCES

84. Boulding, K. E., 'General system theory – the skeleton of science', *General Systems* I, 1956.
85. Boiten, R. G., 'Cybernetica en samenleving', in: *Arbeid op de tweesprong*, The Hague 1965.

86. Verveen, A. A., *Op speurtocht naar processen*, Leiden 1968.
87. Peursen, C. A. van, C. P. Bertels and D. Nauta, *Informatie*, Utrecht 1968.
88. Porter, A., *Cybernetics simplified*, Oxford 1969.
89. Wiener, N., *Cybernetics*, New York 1948.
90. Ashby, W. R., *An introduction to cybernetics*, New York 1956.
91. Shannon, C. E. and W. Weaver, *The mathematical theory of communication*, Urbana 1949.
92. Manescu, M., 'Cybernetic concepts and techniques in the prognosis of the social and economic development of Romania', *Proceedings second conference of the WOGSC*, Oxford 1972.
93. NEN 3009, *Nomenclatuur en definities in de techniek van het automatisch regelen*, May 1958.
94. Nauta Lemke, H. R. van, et al., *Regeltechnische monografieën I*, Rotterdam 1968.
95. Di Stefano, J. J., A. R. Stubberud and I. J. Williams, *Theory and problems of feedback and control systems*, Los Angeles 1967.
96. Sommerhoff, G., *Analytical biology*, Oxford 1950.
97. Conant, R. C. and W. R. Ashby, 'Every good regulator of a system must be a model of that system', *Int. Jour. Sy. Sci.* I, no. 2, pp. 89-97.
98. Mirow, H. M., *Kybernetik*, Wiesbaden 1969.
99. Boiten, R. G. and F. Reimert, *Inleiding tot de regeltechniek*, syllabus Delft University of Technology, 1966.
100. Watzlawick, P., J. H. Beavin and D. D. Jackson, *Pragmatics of human communication*, New York 1967.
101. Johnson, R. A., F. E. Kast and J. E. Rozenzweig, *The theory and management of systems*, 2nd ed., New York 1967.
102. Ulrich, H., *Die Unternehmung als produktives sociales System*, St. Gallen 1968'

BIBLIOGRAPHY *(Authors, listed alphabetically)*

Ackoff, R. L., 'Systems, organizations and interdisciplinary research', *General Systems* V, 1960.

Ackoff, R. L., 'Towards a system of systems concepts', *Man. Sci* . 17, July 1971.

Allen, R. G. D., *Macro-economic theory*, New York 1968.

Angyal, A., 'A logic of systems', in: F. E. Emery (ed.), *Systems thinking* Harmondsworth 1969.

Apostel, L., 'Towards a formal study of models in the non-formal sciences', *Synthese* 12, 1960, pp. 125-161.

Ashby, W. R., *Design for a brain*, London 1954.

Ashby, W. R., *An introduction to cybernetics*, New York 1956.

Beer, S., *Cybernetics and management*, London 1959.

Bertalanffy, L. von, *Kritische Theorie der Formbildung*, Berlin, Borntraeger 1928.

Bertalanffy, L. von, *Theoretische Biologie* I + II, Berlin, Borntraeger, 1932, 1942.

Bertalanffy, L. von, 'The theory of open systems in physics and biology', *Science* 111, 1950.

Bertalanffy, L. von, *Biophysik des Fliessgleichgewichts*, Brunswick, Vieweg 1953.

Bertalanffy, L. von, 'General system theory', *General systems* I, 1956.

Bertalanffy, L. von, *General system theory*, New York 1968.

Bertels, C. P., 'Het model op de wip', *Intermediair* 8, no. 11, pp. 23-25.

Bertels, C. P. and D. Nauta, *Inleiding tot het modelbegrip*, Bussum 1969.

Boiten, R. G., 'Cybernetica en samenleving', in: *Arbeid op de tweesprong*, The Hague 1965.

Boiten, R. G. and F. Reimert, *Inleiding tot de regeltechniek*, syllabus Delft University of Technology, 1966.

Boulding, K. E., 'General system theory – the skeleton of science', *General Systems* I, 1956.

Brillouin, L., 'Life, thermodynamics and cybernetics, ch. 18, in: W. Buckley (ed.), *Modern systems research for the behavioral scientist*, Chicago 1968.

Brönimann, C., *Aufbau und Beurteilung des Kommunikationssystems von Unternehmungen*, Berne, Stuttgart 1970.

Cannon, W. B., 'Organization for physiological homeostasis', *Physiological Review* 9, 1929.

Cannon, W. B., *The wisdom of the body*, New York 1932.

Cherry, C., *On human communication*, 2nd ed., London 1961.

Churchman, C. W., et al., *Introduction to operations research* I, New York 1957.

Coenenberg, A. C., *Die Kommunikation in der Unternehmung*, Wiesbaden 1966.

Conant, R. C. and W. R. Ashby, 'Every good regulator of a system must be a model of that system', *Int. Jour. of Sy. Sci.* I, no. 2, pp. 89-97.

Diesing, P., *Patterns of discovery in the social sciences*, Chicago 1971.

Di Stefano, J. J., A. R. Stubberud and I. J. Williams, *Theory and problems of feedback and control systems*, Los Angeles 1967.

Emery, F. E. and E. L. Trist, 'Socio-technical systems', in: F. E. Emery (ed.), *Systems thinking*, Harmondsworth 1969.

Emery, F. E. and E. L. Trist, 'The causal texture of organizational environments', in: F. E. Emery (ed.), *Systems thinking*, Harmondsworth 1969.

Forrester, J. W., *Industrial dynamics*, Boston 1961.

Hall, A. D. and R. E. Fagen, 'Definition of a system', *General Systems* I, 1956.

Hanken, A. F. G. and H. A. Reuver, *Inleiding tot de systeemleer*, Leiden 1973.

Johnson, R. A., F. E. Kast and J. E. Rozenzweig, *The theory and management of systems*, 2nd ed., New York 1967.

Katz, D and R. L. Kahn, 'Common characteristics of open systems', in: F. E. Emery (ed.), *Systems thinking*, Harmondsworth 1969.

Keuning, D., *Algemene systeemtheorie, systeembenadering en organisatiekunde*, Leiden 1973.

Klaus, G., *Wörterbuch der Kybernetik* I + II, 1965.

Klir, J. and M. Valach, *Cybernetic modelling*, London 1965.

Köhler, W., *Die physischen Gestalten in Ruhe und im stationären Zustand*, Erlangen 1924.

Köhler, W., 'Zum Problem der Regulation', *Roux's Arch.*, 112, 1927.

Leeuw, A. C. J. de, *Systeemleer* I, syllabus Eindhoven University of Technology 1971.

Leeuw, A. C. J. de and W. Monhemius, *Methodologie en inleiding systeemleer* II, syllabus Eindhoven University of Technology, 1973.

Lier, J. J. C. van, *Inleiding tot de thermodynamica*, syllabus Delft University of Technology, 1966.

Lotka, A. J., *Elements of physical biology*, New York, Dover (1925), 1956.

Maarschalk, C. C. D., 'The use of aspect systems in a general model for organizational structure and organizational control', in: B. van Rootselaar (ed.), *Annals of systems research*, vol. 1, Leiden 1971.

Manescu, M., 'Cybernetic concepts and techniques in the prognosis of the social and economic development of Romania', *Proceedings of the conference of the WOGSC*, Oxford 1972.

Meadows, D., et al., *The limits to growth, a report for the Club of Rome project on the predicament of mankind*, New York 1972.

Mesarovic, M. D., D. Macko and Y. Takahara, *Theory of hierarchical, multilevel systems*, New York 1970.

Mihram, G. A., 'The modelling process', *IEEE transactions on systems, man and cybernetics* 2, no. 5, Novembre 1972, pp. 621-629.

Miller, E. J. and A. K. Rice, *Systems of organizations*, London 1967.

Mirow, H. M., *Kybernetik*, Wiesbaden 1969.

Murdick, R. G. and J. E. Ross, *Information systems for modern management*, Englewood Cliffs, N.J., 1971.

Nauta Lemke, H. R. van, et al., *Regeltechnische monografieën* I, Rotterdam 1968.

NEN 3009, *Nomenclatuur en definities in de techniek van het automatisch regelen*, May 1958.

Peters, J., *Einführung in die allgemeine Informationstheorie*, Berlin 1967.

Peursen, C. A. van, C. P. Bertels and D. Nauta, *Informatie*, Utrecht 1968.

Porter, A., *Cybernetics simplified*, Oxford 1969.

Raymond, R. C., 'Communication, entropy and life', ch. 19, in: W. Buckley (ed.), *Modern systems research for the behavioral scientist*, Chicago 1968.

Rosenblueth, A. and N. Wiener, 'The role of models in science', *Phil. of Sci.*, 12, 1945.

Samuelson, P. A., *Economics*, New York 1967.

Schrödinger, E., 'Order, disorder and entropy,' ch. 17, in: W. Buckley (ed.), *Modern systems research for the behavioral scientist*, Chicago 1968.

Schützenberger, M. D., 'A tentative classification of goal seeking behaviours', in: F. E. Emery (ed.), *Systems thinking*, Harmondsworth 1969.

Shannon, C. E., 'The mathematical theory of communication', *Bell Systems Technical Journal* 27, 1948.

Shannon, C. E. and W. Weaver, *The mathematical theory of communication*, Urbana 1949.

Shchedrovitzky, G. P., 'Methodological problems of system research, *General Systems* XI, 1966.

Simon, H. A., *Models of man*, New York 1957.

Sommerhoff, G., *Analytical biology*, Oxford 1950.

Szilard, L., 'Über die Entropieverminderung in einem thermodynamischen System bei Eingriffen intelligenter Wesen', *Z. für Physik*, 1929, p. 840.

Ulrich, H., *Die Unternehmung als produktives sociales System*, St. Gallen 1968.

Verveen, A. A., *Op speurtocht naar processen*, Leiden 1968.

Watzlawick, P., J. H. Beavin and D. D. Jackson, *Pragmatics of human communication*, New York 1967.
Webster's Collegiate Dictionary, 1970.
Wertheimer, M., 'Untersuchungen zur Lehre von der Gestalt', *Psychol. Forsch.* 4, 1923.
Whitehead, A. N., *Science and the modern World*, Lowell Lectures 1925, New York 1953.
Wiener, N., *Cybernetics*, New York 1948.
Wiener, N., *The human use of human beings*, New York 1954.
World Organization for General Systems and Cybernetics WOGSC, *Proceedings of the first and second conference*, 1969/1972, J. Rose, ed.

Yearbooks of the Society for General Systems Research, *General Systems* I-XVIII, 1956-1973.
Young, O. R., 'A survey of general systems theory', *General Systems* IX, 1964.

Zadeh, L. A. and C. A. Desoer, *Linear system theory*, New York 1963.